Better than ObamaCare
(A Free Market Solution to Healthcare)

Better than ObamaCare
(A Free Market Solution to Healthcare)

Nelson A. Paguyo, MD

ISBN 978-1-300-87532-1

Cover designed by Temay R. Paguyo Broadway

To my loving wife, Ester, who has always been understanding and supportive of my professional work and community activities.

To my granddaughter, Madeline Taylor, who was age fourteen years old when she passed away six years ago; who inspired and opened my mind to realize the urgent need to write about and change completely our current healthcare system to be the very best in the world—I especially dedicate this work.

While it is true our healthcare is expensive, full of faults, wasteful, and had been a total failure in providing healthcare for all Americans, it has one redeeming quality that is not found in others: it can deliver immediately the most advanced medical care in the world whatever, wherever, and whenever it is urgently needed in the United States of America, and possibly any where in the world effectively—with minimal delay.

—The Author

CONTENTS

PREFACE

"I turn around and look at the abundance of the American Healthcare System compared to other healthcare systems. I have no doubt that we can cover the entire population of the United States with high quality comprehensive healthcare. The question is whether we have the will and energy to figure out how to do it."

—Elliott S. Fisher, MD, MPH
Medical Research Scientist
Dartmouth Medical School

No American should need to worry about her/his healthcare, especially our most vulnerable citizens—seniors, disabled, unemployed and under-employed, under privileged poor, chronically sick, mentally ill, drug addicts, and alcoholics. Access to our healthcare must be unconditional to all United States citizens, legal permanent residents, and refugees. It is in this spirit that this book was written, and the concepts hereby presented. With good luck, this proposal will act as a catalyst and framework for a national dialogue that leads to necessary healthcare reform.

WEBSTER'S New Universal Unabridged Dictionary defines *crisis* in the following:

- *A stage in a sequence of events at which the trend of all future events, esp. for better or for worse, is determined; turning point.*

- *A condition of instability or danger, as in social, economic, political, or international affairs, leading to a decisive change.*

Is there a healthcare crisis in the United States of America? Consider recent developments.

- When over 16 percent of the population are uninsured and have no means of accessing the healthcare system in spite the federal government's yearly increases in Medicare and Medicaid funding—is that crisis?

- When the cost of healthcare is steadily rising at least twice and at times four times above the inflation rate, subjecting most Americans to crippling financial effects, year after year—is that crisis?

- When 50 percent of all small- and medium-size companies can no longer afford the healthcare benefits of their employees, and large corporations are having difficulties meeting the expenses of employees' healthcare coverage—is that crisis?

- When GM and other large manufacturing, retail, and service companies—because of healthcare expenses— resort to bankruptcy and/or layoff tens and even hundreds of thousands of employees to stay financially afloat—is that crisis?

- When products from the United States are no longer able to compete in the global market because of higher production costs aggravated by healthcare expenses—is that crisis?

- When companies go to outsourcing jobs to developing countries to escape the high cost of labor and healthcare—is that crisis?

- When tens of billions of healthcare dollars are spent for the medical care of millions of illegal immigrants in the United States—is that crisis?

- When big healthcare companies are focused and more interested in the company's bottom line than in patient care—is that crisis?

- When CEOs of healthcare companies are earning upwards of $100,000,000 a year and their management teams not far behind—is that crisis?

- When many Americans are forced, to decide whether to eat or buy their medicine; struggle whether to take the appropriate doses, cut down or stop their prescribed medications to save—is that crisis?

- When more than 50 percent of bankruptcy filings in America are from middle class families who can not pay big medical bills—is that crisis?

These are just a few interesting questions that by themselves may not be critical in causing healthcare crisis, but when taken as a whole the effects are tantamount to a major crisis that pervades every segment of our society, and our way of life. Correcting the ever increasing healthcare spending then becomes a priority of every American (especially so of our elected government officials), for if allowed to continue will inevitably bankrupt the healthcare system, if not our nation.

ACKNOWLEDGEMENTS

I give my sincerest thanks to Al and Carol Dornisch, Darryl Jackson and Jann Nelson, and Jose Valera, J.D., who gave me their constructive views and painstakingly edited the original manuscript of *A Framework for a Universal Health-Care Plan* on which this book was primarily based.

To my two daughters, Liza P. Steman and Alyce P. Muenchow, who both reviewed the original material and gave me feedbacks of their own experiences with the healthcare system, frequently prodding and encouraging me to write it into a book to reach a wider audience.

I specially give recognition to my colleagues Gayl Gustafson, MD, Sharmishtha Raiker, MD, and Lourdes de Jesus, MNS, BNS, who generously offered me their helpful insights.

Many thanks to Loretta Krier, LPN, Marcelle Triantafilou, RN, and Lynn Zhulkie, RN, who helped research the project, and to all the nurses and staffs at HealthPartners Riverside Clinic, Adult Medicine Section, who took the time to read the original thesis at home and gave me their personal perspectives on the problems of our healthcare system.

My deepest gratitude to Araceli Carbonell, who distributed copies of the original manuscript at the 2004 Republican Convention in New York City; Anita Sese Schon, Zenaida Dacanay-Lerma, Erik F. Peterson, and the cadres of ordinary Americans who believed in the idea, and who will advocate and help push healthcare reform to completion.

And last but not least, I acknowledge and very much appreciate the understanding and wisdom of a dear friend and classmate, Teresita Ferrer Estoye, MD, who honestly critiqued the first draft that forced me to carefully rethink this through and work harder to come up with a practical proposal, and a game plan that hopefully will bring healthcare reform to realization.

Again, thank you all!

A NOTE TO THE READER

Dear Reader,

Thank you for your interest. I trust this book is as informative as you expect it to be. Your wish to learn more and bring changes to our present medical care is both commendable and admirable. Welcome aboard!

The section on OBAMACARE gives you a better understanding on the *PATIENT PROTECTION AND AFFORDABLE CARE ACT (HR3200) of 2010.*

My initial attempt to introduce reforms in the healthcare system of the United State of America was not so encouraging. Writing this book is a change in strategy on my part. Hopefully, I'll be able to reach a critical number of Americans who are concerned with the state of our medical care, and are willing to advocate for a healthcare plan that is universal and within our means.

After reading this healthcare initiative and you find the general principles agreeable, I urge that you assist in promoting a healthcare system that is easily accessible and user friendly; giving uninterrupted healthcare coverage to all Americans.

The treatise offers rather radical and comprehensive solutions to our ailing healthcare. Your time, effort, and support are urgently requested. Please direct your campaign to all your family members, friends, acquaintances, co-employees, and especially to your elected senators and congressmen in the U.S. Congress. Encourage them to enact laws creating a universal healthcare plan for America.

Unfortunately, the recently passed HR3200 healthcare bill did not properly address the major issues our healthcare system has. Many consider ObamaCare made the present healthcare worse and more expensive. Although unlikely to happen, I wish President Obama recommends the repeal of HR3200 and have Congress start all over to incorporate the major principles I have discussed.

Let us all act and pressure Washington to fundamentally change our healthcare system before the end of President Obama's second term.

The best campaign strategy is by word of mouth. After you read this book please share it with your friends and relative, and encourage them to call Washington, D.C.

Thank you very much for helping. May our endeavors bring this health initiative to fruition.

Sincerely,

Nelson A. Paguyo, MD

AUTHOR'S COMMENTS

With Americans' determination and demand for better healthcare, and the willingness, cooperation, courage, and the political will of our leaders in government and the private sector to reform this healthcare system, I have no doubt Americans will have a Universal Healthcare Plan soon on the premise of this proposal.

It is my hope that every American recognizes and clearly understands the serious predicament with which our healthcare system is threatened. Allowed to continue on its present course, the current healthcare system of this country will financially go bankrupt within the next decade or so. The implication and/or ramification are disruption in the well-being of our citizens and economic stability of our nation. This proposal attempts to identify, and offers solutions to, the existing core problems of today's medical care and delivery system.

The current healthcare debates, and attempts by the U.S. Congress to reform the U.S. healthcare system, sadly, is based on political games our elected officials play with complete disregard on what the American people really want, and the opportunity to change the present healthcare into a truly effective, efficient and affordable system is lost with the passage of ObamaCare.

One important ingredient that needs to be acknowledged, however, is for all Americans to claim ownership of a United States of America Universal Healthcare Plan if legislated and adopted. It has to be because sooner or later everyone in her/his lifetime will use and become the beneficiary of the system. Acceptance of this important principle means every American, especially the rich, famous, influential, businessperson, politician, and lobbyist *must forego* their immediate or short-term goals in favor of a viable national healthcare plan.

Think of our future as individual Americans and as a nation.

Please remember that no matter how good your healthcare coverage is today, in the not so distant future when you are retired, become unemployed or disabled, sick, and old, expect your good healthcare coverage (today) to suddenly and unexpectedly disappear, leaving you with inadequate, expensive, or no coverage at a time when you most need it. The other consideration one must always keep in mind is some of your close relatives—parents, brothers and sisters, children, and grandchildren, etcetera, may have inadequate or no healthcare insurance coverage they deserve. This is by no means intended to frighten anyone, rather it is a restatement of facts.

Should the reader find areas of this plan controversial, disagreeable, or unacceptable, the author requests that she/he does not withhold her/his support to this health initiative. Rather, the reader is encouraged to discuss it with her/his elected U.S. and State legislators, so her/his concern is properly addressed and reflected in the final legislative draft for the United States of America Universal Healthcare Plan.

I invite the public, who supports this proposal to participate in convincing the U.S. Congress to pass a national healthcare plan. Let's all forget our political affiliations and work united as one for a Universal Healthcare Plan for America.

Let's not delegate our individual responsibility to the next person, or to the next generation. Do whatever you can. Do not underestimate your personal conviction and contribution. It will make the difference in this campaign.

WHAT U.S. HEALTHCARE SHOULD BE

A National Healthcare Plan for all Americans does not have to be like the recently passed Patient Protection and Affordable Care Act, popularly known as the OBAMACARE, or those of Canada, United Kingdom or any European and Asian health systems. With Medicare and Medicaid expenses rapidly increasing and representing significant portions of U.S. budget, our healthcare system has to be reformed.

The U.S. healthcare does not have to be government-controlled, or run as a single-payer.

It is universal in scope that provides a comprehensive medical and dental healthcare coverage for all Americans, legal permanent residents, and refugees.

It can be uniquely an American answer to some of the most basic healthcare concerns we have—accessibility by all Americans; portability; affordability; freedom to choose one's provider and for the provider to practice his/her healthcare profession without interference from others.

Free market principles, especially competitive bidding must be an essential part and strictly enforced component of the healthcare scheme.

Last but not least, the U.S. healthcare system must be simple, easy, and user-friendly for all participants involved at every level of the system from the patients to providers, and healthcare administrators.

I

The United States Universal Healthcare Plan

The Key Features of an Ideal Healthcare Plan

An American Healthcare Plan should have attributes consistent with the best of healthcare schemes. The proposed plan attempts to achieve all the features of an exemplary healthcare system. These are:

- A healthcare system must be simple and easily understood, free of worries, user-friendly, supported by the public, and equally uncomplicated to administer.

- It is affordable and reasonably priced, reflecting the true market value of healthcare products and services in the market place.

- It is portable, readily available, and accessible to all citizens regardless of whether they are in the U.S., foreign country, or anywhere else in the world.

- It provides comprehensive and high quality medical and dental health coverage that easily exceeds the benefits offered by any government or private health program, or any other universal healthcare system in existence around the world today.

- It assures patients the freedom to choose their health providers, easy and universal access to healthcare whenever needed and wherever they may be, and a guarantee of uninterrupted healthcare coverage no matter what circumstances citizens are in.

- It allows healthcare providers unfettered ability and independence to practice their profession.

- It must be market-driven with competitive bidding as the primary method of implementing business transactions at every level of the healthcare delivery system.

- It provides a mechanism for funding health education, and research and development.

- It has mechanisms to establish and finance reserve funds for healthcare within the system.

These qualities of an ideal healthcare plan are important and achievable, if Americans want to seriously reform the present healthcare system into one that is reasonably restructured that prevents abuses, wastefulness, and unsustainable practices. Such a healthcare system is attainable if there is political will in this country devoid of special interest group influences that frequently invalidate good intentions.

Principles in an Ideal Healthcare Plan

To establish an ideal healthcare plan for America, a set of essential principles must be incorporated and adhered to, to make it attainable and realistic.

- The federal government creates a healthcare superfund that will finance the entire United States of America's healthcare system.

- The federal government redistributes the healthcare superfund to the different states and territories of the U.S. proportionate to the per capita population of the states.

- The federal government defines what a comprehensive/basic medical and dental coverage that gives uniformity of coverage, and becomes the reference point from which all competitive bidding for all the citizens' health insurance coverage is based.

- Recipient states are required to purchase health insurance protection for each citizen of the state, or provide the healthcare needs of the citizens through state health programs.

- It permits individuals or state governments to purchase across state lines additional or supplemental health coverage for health products and services that are not included in the basic medical and dental care insurances—including integrative and/or alternative, cosmetics, controversial and unproven experimental forms of treatments, and other forms of non-traditional therapeutic approaches, but are prescribed by the patients' healthcare provider.

- It should not be government-run, with the roles of the federal and state governments strictly limited and clearly defined.

- It should not be a single-payer healthcare system.

- It should be privately administered to promote collaboration between the private/business sectors and the governments.

- It should encourage competition; applying the principles of leverage in a two-step bidding process as an essential part of every business transactions at all levels of the healthcare delivery system—to control effectively the upward pressures in medical and dental care expenditures, and to allow free market forces to determine the true value of healthcare.

- It allows competition among all qualified health insurers across state lines and/or even across countries.

- The winning (health insurance provider) bidder at the state level shall provide the necessary health coverage for all the state citizens designated in the bidding process.

- Enforcement of the state obligation to register and issue healthcare ID cards to all state citizens at point-of-registration.

- It provides a built-in process to finance healthcare education, and research and development both at the federal and state levels of government.

- It has provisions that mandate accumulation of unspent healthcare dollar into healthcare reserves, and to save all excess health dollars; for the purpose of maintaining projected estimates of healthcare expenses, for a period of 10 years in both federal and state reserves.

- It has provisions to establish a healthcare Advisory Council that examines and addresses significant problems encountered, and for the Advisory Council to recommend timely solutions on how to resolve specific healthcare issues and concerns recognized later in the future.

- Controversial medical treatments with political undertones must be delegated to a team of local/state practicing physicians in the private sector, medical ethicists, consumer/community advocates, and state government representatives to decide, and such decisions recommended to the patient's primary physician or healthcare provider.

- Lastly, the winning healthcare insurer must be rendered viable by obligating such insurance company to have reinsurance healthcare policy at all times from any big, stable, and reliable reinsurance company.

These principles serve as guidelines in dealing with the many complex problems facing the health system. Unfortunately, the features of an ideal health plan are not found in a single existing health plan, but rather are in many different healthcare plans.

In an attempt to formulate a healthcare model consistent with the characteristics of some of the best available systems rolled into one, this proposed U.S. healthcare approach has incorporated in it all the qualities of an ideal healthcare plan. To accomplish and develop such a healthcare scheme means, the complete dismantling and rebuilding from scratch the present American healthcare system to give birth to an ideal new system that is universal, affordable, and sustainable.

Reasons Behind the Principles

General principles serve as guidelines in the design of any policy that is adaptable to meet the needs of the people. Such principles must be well understood by everyone if the people have to support and endorse it.

- Simplicity of the healthcare system must be a priority. Most medical insurance contracts, instructions, benefits, and so on are written in the languages of lawyers and business administrators; it is often confusing and difficult to understand by ordinary citizens. Frequently, people need experts to explain the insurance contracts, medical instructions, and benefit literatures, and accept without question what their translators and insurance agents usually tell them. Sometimes agents misrepresent their insurance products that it is not uncommon to hear such misrepresentation and such person was not properly covered in time of their needs. Others just completely opt out of the healthcare system in spite the availability and their eligibility for health coverage, because of the person's lack of understanding of eligibility requirements; medical benefits, and complex enrollee responsibilities.

- The creation of a superfund provides an avenue for providing health coverage to all Americans without additional cost; with pre-existing illnesses of one individual citizen diluted and spread out to a large group gives the insurer a sensible risk at a lower cost.

- The pooling of all the monies earmarked for healthcare enables the federal government to create a super healthcare fund with enormous dollar amount and significance, that even though it is later subdivided and distributed to the different states and territories, each state will still have big enough amount of healthcare money that can not be easily ignored by the business communities; giving the state or territory great advantage and leverage in negotiating realistic price and true value of the health services and products.

- To avoid the inefficiencies and limitations of a universal healthcare plan totally controlled by the government. The role of the government must be restricted to a specific function that encourages private enterprises, and the free market determines the true value of healthcare services and products.

  Universal healthcare plans like those of Canada, England, and Sweden are single-payer and government run. Other European and Asian universal healthcare schemes are mostly government controlled that permit, groups of government-approved private insurances to cover and administer the healthcare services to the population. In these health systems, the services are allocated and rationed because of government restrictions and regulations. Such government control leads to limited patient [coverage] benefits, health equipment, facilities, providers, and other healthcare services. Undue long waiting periods often endanger patients lives; some die before their health providers see them.

  Like in the U.S., healthcare cost and public dissatisfaction had increased to the point that in many countries, private insurances, hospitals, and clinics were slowly introduced and allowed to supplement the healthcare services in their systems.

- The compulsory requirement of the state or territory to use the competitive two-step bidding process in the procurement of health coverage for each citizen, and the employment of the same bidding practice by the winning health insurance company, applied at every level of the healthcare delivery system is aimed at drastically reducing the cost of healthcare.

It is well documented that healthcare products—from health insurance premiums to prescription drug prices; general and specialty hospitals; medical and dental clinics; primary and specialty doctors fees; laboratory and diagnostic groups; physical therapy and rehabilitation centers; nursing homes and hospice facilities; patient homecare companies; long-term care institutions; drug stores and other healthcare charges— widely vary from as low as 50 percent to as high as 800 percent from the baseline price available in a community. These are acknowledged observations that imply abusive overcharging or price gouging by many health insurances; bidding practices can control healthcare product manufacturers and/or providers of healthcare service. Prevention of overcharging and "gaming" the healthcare system must be a priority. Laws against over pricing, abusive and price gauging should be enforced and appropriately punishable.

- The federal mandate requiring buyers and sellers of healthcare services or products the two-step bidding process (across stateliness) at every level of the healthcare system—from buying insurance health coverage: hospital, medical and dental clinics, diagnostic centers, rehab centers, integrative clinics, and other health contracts; medical, pharmaceutical and other medical and/or healthcare products is an important stipulation for the proposed plan to cut effectively medical cost.

Buyers at every level of the healthcare system bargain from a position of strength, and sellers in a compulsory two-step bidding process participate with the primary purpose of winning the contracts that represent sizable business for the sellers; a sure way to drive down the cost of healthcare at sensible and inexpensive level; with the possibilities of significant savings as a feature missing in the present and other health systems.

- Many health insurance companies (usually the smaller non-profits) that provide similar comprehensive coverage are cheaper by 50 percent to over 300 percent less than what other bigger for-profits health insurance companies charge. The two-step bidding requirement across every level of the healthcare delivery system truly offers an effective measure to lower the cost of healthcare in this country.

To illustrate, in 2004, a profitable Minnesota HMO spent an average of $2,400 for each enrollee in their healthcare insurance products. Compare this with the U.S. per capita healthcare expenditure in 2004 of $6,280. The Minnesota HMO is 62 percent cheaper than the national average. Other Minnesota HMOs are cheaper by 30 percent to 40 percent of the national average. Similarly, in 2004 French had a per capita healthcare expenditure of $3,776 or 40 percent lower than the U.S.

- The use of the two-step bidding competition among businesses is a proven method of controlling cost, and definitely will give the American public the true market value of healthcare in this country. Competitive bidding is a principle widely used by many big corporations including Wal-Mart.

- Supplemental healthcare insurance policies must be permissible for healthcare consumers, to purchase added health protection for themselves and their families against healthcare expenses that are not covered with the national healthcare plan. This is similar in principle to Medicare supplemental policy for medical and dental services not covered by Medicare. This may include new, expensive, and experimental drugs, diagnostic procedures, hospice and long-term care, home care, and others already mentioned.

- The registration and issuance of a U.S. national healthcare ID by the state or territory, facilitates the efficient access of the American healthcare system by every American citizen. ID healthcare card is widely used in countries with universal healthcare, especially in Europe. It is user-friendly that all a patient has to do is present the healthcare ID card to a health provider of the patient's choice anywhere in the U.S. (and perhaps in any country), and everything is automatically taken care. As to concerns regarding privacy issues and identity theft, the healthcare ID card is tamper-proof and contains only enough personal information, similarly found for credit cards to conduct health-related business operations, and clinical data systems management purposes.

  The national healthcare ID card stops time-consuming duplication and eliminates gathering of personal information every time health providers are visited. It conveniently avails digitalized electronic medical records, and facilitates the transfer and availability of clinical records for healthcare providers in the clinics or hospitals; consultants; laboratories or diagnostic centers; physical and rehabilitation facilities; and others that need patients' medical records. In addition, the healthcare ID card will also help in the expedient processing or handling of claims related in the business aspects of healthcare practice; making the healthcare system more productive and operationally efficient.

- The provision for healthcare education both at the federal and state government levels insures the availability of funds for health education of our children. Health education, as a permanent part of the school curricula from elementary to high school grades, will inculcate in the young minds of our children good and proven health practices that certainly will prevent diseases in later lives—probably the best measure for prevention of diseases in the future.

If the healthcare educational fund for the children is more than enough to support the program, some of the monies can be used to fund adult patients' education and coordination of patients' treatments with chronic diseases.

- For the same reason, the provisions for the inclusion of both federal and state funds for medical research and development will provide a continuous source of financing for medical research and development. This will assure the country that R & D shall always be funded and not allowed to regress behind other countries in the world, especially, in times of economic slow down.

- The proviso for the creation of the federal and state healthcare reserve funds financed by the unspent and/or saved monies from the healthcare funds and allowed to accumulate are insurances for a well-funded American healthcare system in time of American or global economic disasters. Such a provision in the healthcare law will force the federal or state governments to save and deposit in the reserve accounts monies not spend during economically prosperous years.

The required 10-year projection of estimated healthcare expenses saved and retained in the Federal, and State Healthcare Reserve Funds was designed to ensure funding of any short falls in revenues during an economic crisis—which may last for several years.

- The stipulation to identify and properly study problems encountered by the present or future healthcare system will enable the national healthcare system the necessary course of action to resolve problems that threaten the financial stability of the American healthcare system long before such predicaments occur.

- The acceptance of ownership of the proposed healthcare system by the American people will hopefully make the American people use their healthcare benefits in a reasonable and responsible way, and not abuse or over-utilize such a health plan.

- Controversial medical issues such as abortion, use of experimental healthcare diagnostic, therapeutic managements, withdrawal of treatment, and set standards in the treatment of the elderly and terminally ill patients, must be left to the individual state to draw up policy for the citizens of the state. This will allow the states to draft general policy guidelines for controversial medical treatments best suited for the local residences of the state and the local healthcare providers in the communities to decide and implement.

- And last but not least is the mandate for a healthcare insurance provider to have a reinsurance policy with single and/or multiple large stable reinsurance company that will insure the financial solvency and stability of the primary healthcare insurance company from failure to execute its responsibilities or bankruptcy.

Funding for the U.S. Universal Healthcare System

The federal government creates a superfund by pooling all the monies earmarked for healthcare expenditures. The funding shall come from the same sources as the present healthcare system. The numbers below are from the year 2010.

- Of the total healthcare expenditures to this country, private employers contribute 39 percent or $975 billion for their employees' healthcare benefits.

- The federal government through its Medicaid, Medicare, Veterans Health Administration, TRICARE, Indian Health Service, Federal Employees Health Benefits Program, and the Children's Health Insurance Program

(CHIP) contributes 35 percent or $875 billion to the total dollars spent on healthcare.

- States contribute 12 percent or $300 billion to the national health expenses.

- The out-of-pocket expenses from individuals and families account for 14 percent or $350 billion of the entire healthcare.

If all these available healthcare monies are pooled as a healthcare superfund, the resulting fund will be more than sufficient to provide healthcare coverage for all Americans as described in this healthcare initiative.

As an example, the total national healthcare expenditure in 2010 is $2.52 trillion or $8,129.03 per capita based on 310 million U.S. population.

FICA tax rate is 13.3 percent. It has two components: the Social Security portion is 10.4 percent and the Medicare part is 2.9 percent. ObamaCare increased the Medicare portion of FICA to 3.8 percent. They should be separated. In most universal healthcare models reviewed here, the employees and employers each contribute to the employees' healthcare coverage from 9.5 percent to 20 percent of employee's income. In HR3200 *Sec. 412* mandates a non-electing employer to pay 8 percent of the employee's salary for the employee's healthcare coverage.

This does not have to be the case in America; however, the employees and employers contribution should be higher as determined by actuarial estimates to make sure the American version of a national healthcare is feasible and sustainable.

Overview of the Proposed Healthcare Reform Plan

The federal government, through the act of the Congress of the Unite States, shall create the *United States National Healthcare Plan (U.S.NHP)* or *United States Universal Healthcare Plan (U.S.UHP)*.

The Congress of the U.S. shall legislate a healthcare bill that shall consolidate all finances earmarked for healthcare expenditures into a single healthcare superfund; consisting of all federal healthcare expenditures including the federal employees' health benefits, Medicare and Medicaid programs; present state governments' contribution to healthcare for their poor citizens; present employers' contribution to their employees' healthcare benefits; and employees and individuals' contribution for their own healthcare coverage shall all be thrown into the healthcare superfund for the sole purpose of financing the healthcare of all Americans.

This shall be the basis of an amendment to the U.S. Constitution that the author strongly recommends, called the National Healthcare Reform Act, otherwise, known as the United States Universal Healthcare Plan (U.S.UHP).

Once the healthcare superfund is fully funded through the above-mentioned sources, the federal government then distributes the healthcare superfund monies to all the states and territories of the U.S. on per capita bases. This is the only strategy to make sure all Americans, legal permanent residents and refugees, receive equal allocations for their healthcare. Every recipient states are obligated to use the healthcare superfund money received for the exclusive purpose of acquiring health insurance coverage for each state citizen in the open market.

A two-step bidding process shall be the rule for all healthcare purchases and other business transactions, at every level of the healthcare delivery system from the acquisition of healthcare insurance coverage for the state citizens to the various negotiated contracts for healthcare services and products.

Provisions to allow this plan become sustainable in perpetuity, without additional taxpayers' money, shall be incorporated in the United States of America Universal Healthcare Plan laws.

The primary objective is to design a healthcare scheme that satisfies the criteria of an ideal healthcare plan as stated. The secondary objective is to address and possibly resolve each major problem of concern in the present healthcare system. Examples are escalating healthcare cost, 16 percent of Americans without health coverage, long waiting periods, rationed healthcare, freedom to choose providers, and to practice one's profession, etcetera.

Such a healthcare is possible, and can be achieved by firmly focusing on the principles that help make healthcare affordable at a reasonable cost, universally available and easily accessible by all

Americans; recognizing that the freedom to choose their providers and the ability to practice one's health profession is an undeniable right of the American people.

American public concerns, such as healthcare malpractice, tort reforms, pre-existing health conditions, nationwide health insurance competition, affordable healthcare fees, health prevention, and many more should be addressed aggressively by the health plan. It should seek the proper solutions to the present healthcare problems.

By the very nature of the proposed healthcare, most of the major difficulties of the present healthcare system are directly and indirectly dealt with and resolved.

Only when the ideals of a national healthcare system are adopted will there be acceptance by the American public to reform this health system. The U.S. Congress can act into law all the features of an ideal health system only if and when elected government officials will put the interest of the American people and our country ahead of interest-groups, self-interest, and the political games our elected officials play in Washington, DC.

II

Creation of a Universal Healthcare Plan for America

"We, the People of the United States, in order to form a more perfect union, establish justice, insure domestic tranquility, provide for the common defense, *promote the general welfare, and secure the blessings of liberty to ourselves and our posterity,* do ordain and establish this Constitution for the United States of America."

The preamble of the United States Constitution states so eloquently why the federal government should step in to aggressively address, drastically correct, and remedy the many faults of our current healthcare system. Questions are: *Is the United States ready for universal healthcare? Do our elected leaders have the political will and courage to fix and fundamentally change the present healthcare system? Is it possible to fashion an economically affordable universal healthcare plan that can last for many generations?*

It is about time our government institute a realistic and financially feasible healthcare system that covers every Americans with the least of government interferences and restrictions.

In the presidential debate in St. Louis, Missouri on October 17, 2000, President George W. Bush was asked this question: "Would you be open to the ideal of a National Health Plan?"

He replied: *"I'm absolutely opposed to a National Healthcare Plan. I don't want the federal government making decisions for the consumers or providers. I remember what the administration tried to do in 1993. They tried to have a National Healthcare Plan, and fortunately it failed. I trust people; I don't trust the federal government. I don't want the federal government making decision on behalf of everybody."*

President Bush's statement clearly indicated that his opposition to a National Healthcare Plan is not in its concept of universal access; rather the implication of the federal government making the decision for both the patient and the doctor. The universal healthcare system for the United States of America presented here is not government run. It is a market and consumer-driven healthcare system that maximizes the principles of a free market economy.

Around 2006 there was a gathering at the Mayo Clinic in Rochester, Minnesota. Two hundred and fifty policy leaders who are experts in healthcare, business, academia, and government attended the meeting. The subject of the conference was Health System Reform. Its goals were *"to highlight the urgent problems facing the healthcare system and begin developing a new vision of healthcare in America."* Participants included distinguish names in the various fields. Serving as panelists were Stuart Butler, PhD, vice president of domestic and economic policy studies at the *Heritage Foundation*; David Kendall, senior health policy fellow at the *Progressive Policy Institute* in Washington, D.C.; professor emeritus of economics and health research and policy at *Stanford University* in California; Victor Fuchs, PhD and Ezekiel Emmanuel, MD, an oncologist and chair of the Clinical Bioethics Department at the *National Institute of Health*. The symposium was co-chaired by Hugh Smith, MD, former chair of the *Mayo Clinic Rochester* Board of Governors. The consensus among the panelist was *"the nation needs universal healthcare coverage."* Seventy-four percent of the 250 participants strongly endorsed the concept that *"there is a moral imperative for all U.S. citizens to have health insurance."* There was, however, fundamental disagreement among the panelist as to how universal health coverage can be achieved.

The *Citizens' Healthcare Working Group* is a 14-member body of physicians, nurses, hospital officials, economist, and other experts charged by the U.S. Congress to look into how to fix and find out what Americans thinks of the present healthcare system. The U.S. Comptroller General appointed the panel in February

2005. So far the group completed its interim report based on six meetings with experts, thirty-one meetings, and other healthcare events in fifty communities and major healthcare opinion surveys conducted in 2002. Five thousand commentaries from the public were also reviewed. The group's interim report recommended the following:

- o *"It should be public policy that all Americans have affordable health care."* Possible revenue sources mentioned in the report were from income taxes, businesses, sales, and tobacco and alcohol taxes.

- o *"A standard core benefits should be defined, covering physical, mental, and dental health."* This benefit package should include wellness, preventive, acute and chronic medical care, prescription drugs, etcetera, and must be evaluated and updated periodically by an independent, nonpartisan private/public group.

- o *"A national program, private or public, must protect all Americans against high out-of-pocket costs and provide financial protection for low income people."*

- o *"The federal government should direct a national effort to develop and integrated public-private community networks of institutions and healthcare professionals that would provide high quality care to vulnerable populations."*

- o *"Quality of care and efficiency should be encouraged by using federal funds from programs such as Medicare and Medicaid to integrate systems in information technology and electronic record system; fraud and wastefulness, and consumer-friendly data on pricing."*

- o *"The way end-of-life services should be financed and restructured to increase access for people with incurable conditions to services in the setting they choose."*

The United States of America, considered to be the wealthiest nation, also has the most advanced medical science and healthcare system in the world. The cost of healthcare and the premium for medical insurance, however, are becoming prohibitive to the average American. More than 16 percent of our citizens are without medical coverage. Their access to our healthcare is virtually nonexistent. In addition, healthcare expenses are becoming a significant financial burden to American companies affecting their ability to compete in the global economy in a very negative way. This environment of escalating health expenditure together with the inaccessibility of medical care by 47,000,000 Americans dictates a radical restructuring of the present healthcare system, and the development of an economically plausible universal healthcare, is urgently needed.

While it is true that a democratic government like the United States of America can not take responsibility for all the needs of a society, it is, nevertheless, the duty of the government to see to it that it protects and secures the well beings of all its citizens. Governments pass regulations to force industries to improve service and lower cost, and even out the playing fields of industries to a fair and balanced competition. These laws are enacted when industries are out of control to the detriment of the public interest. The healthcare system of this country has fallen into this category. It poses serious difficulties in people's ability to access medical care, America's ability to fairly compete in the world economy, and all the other adverse societal ramifications of a totally dysfunctional American healthcare system.

The United States of America has the Medicare, Medicaid, and other federal healthcare programs for the armed forces, federal employees, Congress, Justice and Executive branches of the federal government that are either free or low cost to the beneficiaries. These federal health programs, however, do not take into consideration the variety of healthcare issues the general public is subjected to, like the sky rocketing cost of healthcare causing millions of our citizens' inability to afford medical care or medical insurance.

A national healthcare plan that covers every American citizen regardless of his or her race, religion, social standing, employment or economic status, and state or country of residence must be unconditional. The magnitude of such a national healthcare plan can only be promulgated and instituted by the federal government.

It has to be legislated by Congress, and signed into law by the president of the United States of America for it to be inclusive and effective. Preferably it should be an Amendment to the U.S. Constitution to prevent easy manipulation or changing at the whims of politicians. It can be named the *United States National Healthcare Plan (U.S.NHP)* or *United States Universal Healthcare Plan (U.S.UHP)*. The latter sounds preferable. It will be the name of the healthcare plan used in this proposition.

The idea of a universal healthcare plan is nothing new. Most European countries and Canada have different and varying versions of the same idea. However, they differ in their approaches in financing and degree of effectuation. Unfortunately, each of these countries' national healthcare system is starting to experience financial pressures similar to those of the U.S. medical care in some countries with state-sponsored healthcare are beginning to be allocated and/or rationed by decreasing their services, and limiting their citizens' access to medical care. In recent years, major economic problems of the various national healthcare plans were emerging. Countries like Canada, England and Sweden began having difficulties finding the right solutions to the ever-growing budgetary shortfall these governments have for their healthcare programs.

The Ideal American Healthcare Plan

American medicine is the best in the world, in its medical practices, availability, and scientific advances. It is, however, generally perceived as a high-priced and inefficient healthcare system. It is riddled with bureaucratic red tape formulated by government administrators, business executives, and other non-medical professionals. A litany of paperwork to comply with both government and private industry regulations render the system highly ineffective and expensive. Navigating the system is time-consuming, complicated, confusing, and almost impossible for the average American to understand, especially the elderly. Medicare is a good example. The present healthcare system is disease-oriented in its management approach and medical procedures financially favored. Prevention and education—universally recognized as the cheapest of medical managements—are proven to extend life and foster good health are not emphasized, and most often overlooked or neglected. With an uninsured population approaching 47,000,000 people, and healthcare rapidly becoming too expensive, Americans can no longer disregard the problems facing their healthcare. We, the people, must demand from our federal government to legislate immediately the *United States Universal Healthcare Plan.*

For the United States of America Universal Healthcare Plan to work effectively, intense scrutiny of the general principles forwarded from which the proposed universal healthcare plan is based upon, be undertaken. It will be to our advantage if there is a national debate before any act of Congress. Emphasis in the various problems of the present healthcare system should be carefully identified, studied, and resolved before the establishment of a universal healthcare plan for the U.S. The universal healthcare plan proposed can serve as a framework or working model for a national forum for healthcare reform. And since this country of ours strongly believe in a free market enterprise with the least interference from the government, our healthcare system must and should be based on the guiding principles of free market economy.

For organizational purposes the U.S. Universal Healthcare Plan (U.S.UHP) can reasonably be a separate section/department of the Department of Health and Human Services. Although the U.S.UHP becomes a section/department of the DHHS, it has to be designated by law as independent from the other functions of the DHHS. It should be autonomous if it is to focus and achieve the objectives for which it was created. It can be under the direct supervision of the Secretary of the DHHS.

The primary goal of the U.S. Universal Healthcare Plan is to provide all the money for the healthcare of all American citizens regardless of race, religion, socio-economic status, and residence. A secondary but very important objective is to help finance medical research and development, and education both in the federal and state levels. *These goals are easily attainable by pooling all the financial resources of the country earmarked for healthcare.*

The stated objectives can be effectively met by forming two integral parts of the U.S. Universal Healthcare Plan—the *Healthcare Security Fund (HSF)* and the *Healthcare Security*

Agency (HSA). The HSF and HSA shall be distinctly separated from one another to clearly delineate their functions.

Outline

I.   Federal Government creates the United States Universal Healthcare Plan (U.S.UHP) consisting of the:

A.   Healthcare Security Fund (HSF): collects all the monies earmarked for healthcare to finance the entire healthcare for all Americans.

B.   Healthcare Security Agency (HSA): distributes the HSF funds to all the states and territories on a per capita basis, and writes healthcare policy guidelines stated as general principles.

- o   HSA shall take 5 percent of the total HSF fund for administrative expenses; health education, and medical research and development.

- o   The writing of healthcare policies shall take into serious consideration, address and reform all the principal problems confronting the present healthcare system, e.g., high cost of healthcare; malpractice and tort reforms; defensive medicine; limited or no compulsory competition among healthcare companies; pre-existing conditions; etcetera.

- o   Unspent funds are saved and deposited as the Healthcare Security Fund-Reserve Fund (HSF-RF).

II.   State Government establishes the State Healthcare Fund (SHF) agency that shall enforce and administer all Federal Healthcare Laws (FHL).

A.   SHF agency shall consist of an administrator (CEO); appointed by the governor of the state and confirmed by the state legislative branch of the state government, and 12 committee members (directors), appointed by the governor and confirmed by the legislative branch of the state government.

B.   SHF agency shall take 5 percent of the total SHF funds for administrative, healthcare education, and medical research and development.

C.   It shall supervise and impose the competitive two-step bidding process across and in all levels of any business transaction, in the healthcare system.

D.   It shall register and issue the National Healthcare ID card to all state citizens.

E.   It shall see to it that all federal health laws are followed.

F.   Unspent funds are saved and deposited as the State Healthcare Reserve Fund (SHRF).

III. Implementation

A.   The HSA shall distribute the HSF funds to the different states and territories of the United States of America on a per capita formula, based of the state's own population census.

B.   SHF Agency buys healthcare coverage for state citizens by conducting the competitive two-step bidding approach for a designated state, city and/or county population from private health insurances, HMOs, PPOs, etcetera.

C.   Winning company provides healthcare coverage for all the stipulated state, city, and/or county resident citizens.

D.   Winning company likewise conducts a two-step bidding competition from healthcare providers and other companies providing health products and services.

E.   County registers and issues a National Healthcare ID card for every state, city, and/or county resident for easy access of healthcare services anywhere in the U.S.

F. State, city, and/or county citizens choose their healthcare provider, and access the healthcare system by presenting their healthcare ID card to the provider.

G. Health providers bill the insurance company that provides coverage for the county citizen.

H. Health education and medical research and development shall be funded from the 5 percent allocated for such purposes.

I. All savings/unspent funds shall be used to finance the State Healthcare Reserve Fund.

J. The winning healthcare insurance company shall be compelled to purchase a re-insurance policy from leading dependable re- insurance companies.

Functions of the Government and Private Sector

Roles of the Federal Government

The only roles the federal government has in the healthcare system of the nation shall be those defined by Congress in the U.S. Universal Healthcare Plan through the specific functions of the Healthcare Security Fund and Healthcare Security Agency.

Healthcare Security Fund

The Healthcare Security Fund shall be empowered to cooperate with the Internal Revenue Service and other federal agencies to collect and consolidate into a single healthcare fund all the monies designated for healthcare expenditures from the various branches of the federal government, state governments, private employers/business institutions, and individual taxpayers of the United States of America. ***This is how the HSF shall be financed.***

- The federal government shall transfer all budgeted monies for Medicare, Medicaid, armed forces, and veterans medical benefit programs, healthcare benefit program for federal employees, the members of the legislative, judicial and executive branches of the federal government, and other federal healthcare entitlement programs into the Healthcare Security Fund.

- The state governments shall be required by the federal government to contribute a state's fair share of financial burden equal to the present amount of money annually designated for a state's medical assistance program, or for the healthcare coverage of the state's poor citizens. Such amount shall be expressed as a fixed percentage of the state's annual revenue. Actuarial analysis of the various states' healthcare expenditures for their poor citizens shall be conducted to determine a percentage of a state's annual revenue appropriate as each state's contribution to the HSF. The median or average percentage (whichever is higher) arrived at shall be equally applied to all the states of the union as the state's contribution to the HSF.

- American employers presently provide substantial financial contributions for the healthcare coverage of their employees. The federal government as the Healthcare Security Fund Tax (HSFT) shall collect that amount. The employer's HSFT shall be uniformly expressed as a constant percentage of either a company's gross revenue, pre-tax profit or the employer's present voluntary contribution to their employee's health benefit. The percentage shall be worked out by actuarial estimates that shall show such percentage can continue to sustain the adequate funding of HSF indefinitely into the future. *For example: large and medium-size companies 12 percent of the total annual employees' salaries is a reasonable contribution; smaller companies with more than 300 employees, 10 percent; small companies having more than 100 employees, 8 percent; the very smallest companies with less than 100 employees, 6 percent, and companies with 20 or less employees, 4 percent. These are arbitrary figures.* The actual numbers have to be determined and shall replace the Medicare portion of the FICA.

- Employees and self-employed individuals shall contribute to the Healthcare Security Fund in the form of an individual Healthcare Security Fund Tax—just like Medicare tax. The employees' HSFT shall be directly deducted from their paychecks. Self-employed individuals shall pay their HSFT quarterly. The employees' and self-employed individuals' HSFT shall be equal to the amount of money each individual employee and self-employed person presently pay or are expected to pay for her/his own and/or family healthcare coverage including out-of-pocket expenses like healthcare and dental insurances and other deductibles. This amount shall be more accurately calculated and determined by actuarial valuation. It shall be uniformly expressed as a percentage of an individual's gross income, singly or jointly filed, and shall be tax-deductible. *Unlike Medicare, the entire individual's gross income shall be subjected to the HSFT.*

  To prevent companies and/or self-employed businesspersons from avoiding to pay their fair share through existing tax loopholes, a minimum HSFT shall be imposed based on the gross income/revenue of the company/person.

- All co-payments collected by healthcare providers at point-of-service shall be remitted to the Healthcare Security Fund.

- Retirees' contribution to the HSF shall be determined as their fair share expressed as a percentage of their total benefit-incomes from social security, private pension, 401k and IRA, and other investments. The percentage of the benefit-income shall be assessed only if the individual's total benefit-income is above poverty. Any amount of the retiree's benefit-income and other investment incomes above poverty shall be treated as ordinary income, subject to the same rate as the employees and self-employed individuals' HSFT as suggested above. Payments presently paid by seniors for their Medicare A, B, and D benefits shall be discontinued.

- It can not be overemphasized that members of our society who are unemployed and/or poor shall not be assessed the Healthcare Security Fund Tax. The unemployed and indigent people shall, however, pay the appropriate co-pays required at the point-of-service. Persons who are truly poor and can not afford the co-pay shall still pay the co-pay necessary at point-of-service but shall be reimbursed by the state with mechanisms to be determined.

- Funds collected from foreign governments or foreign insurance companies for the healthcare expenses of their citizens who are illegal or non-permanent resident/non-refugee aliens in the U.S., and who are in the United States to take advantage of the advanced medical and dental care available in that place, shall be deposited at the HSF.

- Foreigners treated in the U.S. shall be responsible for their healthcare bills and shall be collected and owned by the healthcare providers. The federal government shall establish mechanisms for healthcare providers to collect unpaid bills incurred by foreigners.

Healthcare Security Agency

The main function of the Healthcare Security Agency is to manage, administer, and disburse the Healthcare Security Fund's fund to all the states of the United States of America. A secondary function is to write (healthcare) guidelines stated as general principles for the United States of America Universal Healthcare Plan with the help and collaboration of the ***Department of Health and Human Services, practicing medical experts, consumer advocates and other healthcare policy makers.*** The guidelines shall consist of policy and/or regulatory guidelines that are not restrictive and inflexible, that shall not give the government too much power to control, rendering the healthcare plan inflexible, incompetent, and difficult to administer as seen in other national health systems of other nations.

HSA shall also have the power to establish healthcare plans for all Americans residing abroad, and/or exclude federal personnel preferring to retain their present separate healthcare coverage.

Included also in the federal healthcare guidelines are provisions to prevent monopoly of the healthcare industry by a few dominant healthcare organizations, and mechanisms that will guarantee the solvency of the healthcare into the future.

NO ONE HEALTHCARE COMPANY OR ORGANIZATION SHALL PROVIDE HEALTHCARE COVERAGE TO MORE THAN 10 PERCENT OF THE ENTIRE U.S. POPULATION.

To avoid misunderstanding, clarification of certain words used is in order. The word *county* shall also mean and/or include province, also parish as used by Louisiana. The word *state* herein shall also refer to and include territories and protectorates of the United States of America and all Americans living in foreign countries, and other alternative health plans that may be formed for federal employees, etcetera.

An *American Living Abroad Healthcare Agency (ALAHA)* shall be established for U.S. citizens living abroad. ALAHA shall be responsible for providing healthcare coverage for all Americans residing in foreign countries. The ALAHA shall be funded, treated, and operated just like any state. For an American citizen residing abroad to qualify for the U.S. Universal Healthcare Plan, she/he—if of legal age, their parents—if minor, or their legal guardian—if mentally incompetent, must file an income tax return annually with the United States Internal Revenue Service. ALAHA may conduct registrations of Americans living abroad at American embassies, consulates or military bases. Completed U.S.UHP application forms shall be forwarded to ALAHA central office in Washington, D.C.

Similarly, if the other branches of the federal government, like the armed forces and other federal employees, want to administer their healthcare benefits, healthcare plans similar to ALAHA shall be established, treated and considered like any other state.

The other option is [if these federal entities want to opt-out from the U.S.UHP] to remain separate and retain their present healthcare benefits as presently financed and managed, may do so, and be completely excluded from the U.S.UHP permanently.

Disbursement of the Healthcare Security Fund's fund to the states shall be conducted in the following manner:

- **Five percent of the Healthcare Security Fund's fund (HSF5 percent) shall be retained by the federal government for the following purposes:**

  Ten percent of the HSF5 percent shall be used for administration. Forty-five percent of the HSF5 percent shall be for federally-funded medical research and development programs. Forty-five percent shall be for federally funded medical/health education, particularly for our children from K-12, and other healthcare initiatives.

  Unspent HSF5 percent funds for administration, medical research and development and health education shall be deposited in a *Healthcare Security Fund-Reserve Fund (HSF-RF)*, to be used only as intended above.

- **The remaining 95 percent of the Healthcare Security Fund's fund (HSF95 percent) shall be divided and given to the different states of the United States of America. Distribution by the federal government of the HSF95 percent to each state of the United States shall be dependent on the population of a given state on a per capita basis. The distributed HSF fund to the different states shall be referred to as the State Healthcare Fund (SHF).**

  Each state shall be ordered to form its SHF agency or department before receiving the allocated State Healthcare Fund from the federal government. Part of that mandate is for the state to setup a State Healthcare Reserve Fund (SHRF), funded mainly by any SHF savings or incomes that may be generated.

  Just as important a function as the disbursement of funds, the Healthcare Security Agency shall be empowered to formulate policies and guidelines based on general principles (rather than specific policies) that insure

Healthcare Security Fund's purposes and programs are properly executed and met at the federal and state levels. This, however, shall not mean to regulate, control, dictate and/or interfere in the healthcare affairs of the state and/or of the healthcare industry.

- Healthcare Security Agency shall establish a *Federal Guideline for Healthcare (FGH)* defining and describing the rules how it shall be implemented and type of healthcare coverage each American citizen is entitled to. It shall be enforced uniformly—without exceptions— throughout the United States of America and shall be the basis of healthcare coverage given by private insuring companies and/or by the states if they choose to provide healthcare coverage for their citizens. FGH shall contain comprehensive Basic and Major Catastrophic Medical coverage which shall include, among other things but not limited to: outpatient, emergency and in-hospital patient care, diagnostic workups and prescription drugs, preventive care, mental health and chemical dependency treatment, dental care, physical therapy and rehabilitation, alternative-complimentary or integrative medicine, and long-term nursing home and home care benefits, but not cosmetic surgeries unless medically indicated and determined by independent clinicians. It shall be an improvement of existing guidelines or information describing Basic and Major Health Insurance coverage. An important part of the FGH is to encourage, promote, and develop a good patient-doctor relationship by giving the patient the freedom to choose her/his primary healthcare provider, specialist, clinic, and/or hospital at any time the patient thinks it is necessary in her/his care.

  Supplemental healthcare insurance, at the expense of the private person desiring to have such coverage, shall be allowed to purchase additional coverage for medical and dental services not covered by the U.S.UHP.

The Federal Healthcare Guideline shall be carried out by the states. Each state, however, shall also have the authority to modify and improve the guidelines to better fit and serve the healthcare needs of its citizens.

- A *Healthcare Malpractice Disability Protocol (HMDP)* shall be draw up by the Healthcare Security Agency. It shall follow more or less the general principles adopted and employed in the *Workman's Compensation Disability Rating Protocol (WCDRP)*. Parameters for maximum and partial permanent disability ratings, and pain and suffering compensations shall be established. This shall become the official federal protocol for awarding victims in healthcare injuries and malpractice suits.

- Formulate a fee schedule that is sensible and uniformly applied throughout the United States of America. A logical and available method currently in use is the *Resource Based Relative Value Units* system *or Relative Value Units (RVU)* for short. RVU is a standardized method of compensation that attempts to combine and quantify the effects of the cost of practice, the risk and complexity involved in decision-making, the time spent, and the actual work itself. It is presently and widely used by most manage healthcare organizations.

- Develop an application form for the enrollment of all the residents in all the counties of all the states of the United States of America. The application form shall contain the name of the county and state or U.S. embassy or agency from which it was issued. This shall be called the *U.S. Healthcare Plan Application (U.S.HPA)* form.

It shall contain basic personal information such as: name, gender, marital status, date of birth, place of birth (city, county, state and country), citizenship, residential address, employer—if employed—and any other personal information deemed pertinent and necessary for identification, i.e., social security number and naturalization number, date, and place. *It shall also contain the applicant's recent passport-size picture, fingerprints, and other biometrics of the applicant to prevent fraudulent registration. With the advent of digital facial recognition and/or other biometric technologies, the individual applicant's picture, fingerprints and/or other biometric measurements shall be taken at the county's place of registration; just like how it is done for driver's license or other forms of state issued ID.*

Included in the application form is a sworn statement by the applicant of legal age or by her/his parents or legal guardian, if minor or legally declared mentally incompetent, that all the information given is correct. The information in the application shall be verified by supporting documents submitted and notarized by qualified county employees—preferably with security clearance—who shall be notary publics.

Only U.S. Healthcare Plan Application form bearing the name and issued by the applicant's county and state of residence shall be recognized as the official application form for that county and state of the applicant's point-of-registration or residence.

All the information in the U.S.HPA form shall be clearly typewritten to avoid misreading or misinterpretation of data in the application or better yet the information immediately entered into the computer while applicant is registering. All information written or entered in the computer shall be carefully and doubly rechecked for topographical errors and/or misspelling by both registrant and county staff to insure accuracy.

The U.S.HPA form shall be used as the standard document for registration to avoid variation and confusion. Every state shall be required to use this federally approved form in order for its citizens to be validly considered as participants and enrolled to be covered under/by the United States of America Universal Healthcare Plan.

Once a citizen has completed the registration with the U.S. Universal Healthcare Plan, she/he is issued a U.S.UHP health card, which will enable her/him to access the healthcare system of the U.S. by simply presenting the health card at point-of-service—anywhere and anytime whenever medical services are needed.

All completed individual U.S.HPA forms shall be kept at the county's or state's Bureau of Vital Statistics filed (preferably) with the Birth and Death Certificates, or in a state's computer database if available. Such computer database system shall be backed up by another separate computer or other systems, i.e., digital memory banks, computer external hard drives, flash drives, CDs, digital camera cartridges or the likes, stored elsewhere to make available to the county and/or state duplicate/accurate information for restoration of data in the event database is lost or destroyed.

Roles of the State Government

As directed by the federal government each recipient state of the State Healthcare Fund shall be required to establish an independent state government agency that will administer the SHF. General guidelines on how each state shall setup the SHF agency shall be federally dictated by policies from the Healthcare Security Agency. Once the state has organized the state's SHF agency/department, and is operational the federal government shall immediately transfer the designated state's SHF to the state.

State Healthcare Fund

The State Healthcare Fund agency/department shall administer and manage the funds of the SHF for the sole purpose of accomplishing the objectives of the Healthcare Security Fund.

The agency shall have an administrator and a committee. It shall have an administrator appointed by the state governor and confirmed by the legislative body of the state. She/he shall have a term of office for eight years (or for some other term length at the discretion of the individual state's legislative body). The administrator shall act as the CEO and chair of the committee of the agency. The chair shall cast the deciding vote to break a tie in the committee's deliberation.

The state governor shall also assemble a SHF agency committee at a minimum of once every eight years made of 12 members—equally divided among members of the consumer group, health professionals in active practice, and business leaders. The committee shall play the role of a board of directors of a company.

The SHF agency shall be accountable to the state legislative branch of government. The state legislative body shall oversee but not dictate, control, or interfere in the affairs of the SHF agency.

The committee shall have the following purposes:

- To supervise the SHF agency to insure that the federal mandates and objectives of the U.S.UHP are properly realized and enforced;

- To apply federal healthcare guideline in formulating policies and health initiatives that will best serve the residents of the state;

- To screen and qualify registered health insurance bidders competing for the counties or states' citizens healthcare coverage;

- To verify and secure the best qualified bidder with the lowest bid is selected and contracted;

- To create a State Healthcare Reserve Fund (SHRF), safe guard the financial assets, and seek and/or hire financial advisers in investing the SHRF;

- To do other functions necessary to secure the integrity and functionality of the SHF.

Once the SHF agency has been organized as federally demanded, the federal government shall disburse to the state its State Healthcare Fund. The state then shall execute and budget the fund as directed by the federal government.

Each state shall retain 5 percent of its federally allocated SHF (SHF5 percent) and shall be distributed in the following way:

- Ten percent of the SHF5 percent shall be used for administrative expenses. Forty-five percent of the SHF5 percent shall be spent for state-sponsored medical/health education, particularly for the children from K-12, and healthcare initiatives. The remaining forty-five percent shall be used for medical research and development. Funds not spent annually for administration, medical education, medical research, and development shall be returned to the state SHF agency as the SHRF. Such funds, however, may be given back in the years to come to cover (unforeseen) over-budget expenses.

The remaining 95 percent of the SHF (SHF95%) shall be equally divided among all the residents of the state. This shall

represent the dollar amount for the healthcare coverage of each citizen and legal alien with permanent resident/refugee status in the U.S. residing in the state. This will be referred to as the *Individual Citizen's Healthcare Fund (ICHF)*.

Once the state has received its State Healthcare Fund and has calculated the Individual Citizen's Healthcare Fund, the state shall determine how to provide healthcare coverage for each citizen in the state. There are three possible options the state shall have.

- A state may elect to self-administer the SHF or ICHF itself by a mechanism determined by the SHF agency. This option, however, may present future problems such as graft and corruption, politicizing the healthcare, overregulation by the state, lobbying, and other possible state government activities that render the healthcare system ineffective and inefficient. It is also an accepted fact that the private sector can adapt, act, and adjust much faster to rapid and ever changing business climates especially in the healthcare industry. The state also becomes the single-payer. As such, overwhelming problems had been observed in recent years in many national health systems where single-payer controls almost every aspect of the healthcare system.

  However, in the event the lowest bid is above the ICHF, then, the state may elect or be forced to self-administer the SHF. In this case, any cost overrun shall be the responsibility of the state.

- The second or recommended option is for each state to procure/buy private healthcare insurance coverage for each of its citizens in the county where she/he lives. The state shall offer health insurance companies (HIC), managed-care organizations (MCO) such as the health maintenance organizations (HMO), preferred provider organizations (PPO) and/or other legitimately organized healthcare provider's associations (HPA) to bid for the right to provide healthcare or medical and dental insurance coverage for the entire county population. ***This shall be*** *done through a two-step bidding process—secret and open public—in the state's county government seat.*

- And the third option is for a state to elect combining both first and second options to accommodate counties where bidding is not applicable or appropriate.

The main reason for the two-step bidding process is to advance competition, transparency, and fairness. This primary goal is likely to be achieved in the following manner.

To avoid monopoly, no single HIC, MCO, HMO, PPO, HPA, and/or a consortium of insurance exchange shall be awarded to issue healthcare coverage to no more than 20 percent of a state's entire population. However, this is subjected to the discretion of the SHF agency. Exceptions are states, with less than 500,000 populations. Affiliated, merged companies, consortiums, and exchanges shall be treated as one company or organization.

- Each state shall have the responsibility to advertise widely in the state and nationwide, and do everything possible and practical to support competition before conducting the business of bidding for the healthcare coverage of the entire population of a county. Enough time shall be given for every healthcare organization in the country to register.

- Large counties may be subdivided. An arbitrary number of 250,000 residents may be considered for bidding. Counties with 500,000 or larger population may be divided to two or three groups to inspire competition and prevent monopoly. Smaller counties with less than 20,000 people shall be lumped together to create a larger pool and offered for bidding as one; except in instances where there are smaller healthcare organizations financially qualified, willing, and capable of providing comprehensive healthcare to the community. These options shall be under the discretion of the SHF agency.

- In the event one company consistently presents the lowest bid, and has been found to have already insured 20 percent of the state's population, the state will do one of four things: 1) disqualify the winning bidder; 2) the state chooses to provide healthcare coverage to the county population; 3) the state offers the other bidding companies to insure the rest of the population at the same or lower price of the lowest bid submitted, and 4) redo the bidding process.

- Competing HIC, MCO, HMO, PPO, HPA or consortium of insurances or insurance exchanges shall register with the state to qualify in the bidding process. A registration fee—the amount to be determined—shall be collected to help defray expenses. Only registered bidders shall be allowed to submit their *secret bids* in writing to the state SHF agency. Registered bidders shall be forbidden to communicate, consult, or discuss with each other before and during the bidding period. Failure to comply shall lead to disqualification.

- Immediately after the bidder's registration the SHF agency shall begin accepting written secret bids. The bidders shall be given ample time to analyze their bid. One month (or any amount of time decided as appropriate) after the written secret bidding is closed, SHF agency shall make a public announcement and notify all registered bidders as to when and where the open public bidding shall be conducted.

- One week after the written secret bidding, the SHF agency shall notify all the registered bidders, announce and publish to the general public the lowest bid submitted. **The name of the lowest bidder shall not be revealed.** This will allow other companies enough time to reconsider their submitted secret bids.

- The open public bidding shall be conducted as an open public forum similar to a public auction. It is, however, a *reversed auction*. It shall be directed by the SHF administrator and helped by at least one county official— preferably by the county secretary or registrar officer. The registered bidders, any interested citizen, and community organization or healthcare advocacy group of the county to insure fairness and guarantee the lowest bid shall attend it.

- The reversed public auction shall start with the lowest bid submitted. Only registered bidders shall have the right and opportunity to change their secretly submitted bids and compete.

- Once the bidding process is completed and an official winner is declared, the state shall notify the county of the winning bidder, and encourage the residents of the county to start registrations with the U.S.UHP. A registration fee per applicant may be charged by the county to defray cost.

- Each winning bidder shall be awarded a three- to five-year contract. It shall be required to provide a ***reinsurance policy*** to insure the winning bidder can meet its healthcare financial obligations to all the county residents during the contract period.

- It shall be the responsibility of the state to maintain at all times a State Healthcare Reserve Fund equal to at least the amount of 10 years of SHF. This shall be used to supplement or totally finance the SHF in the event of economic downturns in which the federal government can not come up with enough money to sufficiently finance the healthcare needs of the states.

- The SHRF shall come from the SHF savings, bidder fees, and any other incomes that may be generated. Any excess funds above the projected amount equal to the 10 years of financing the SHF set aside and saved as SHRF shall be used only for state healthcare initiatives, health education, research and development, and nothing else. The state shall be prohibited from using any SHF or SHRF for social programs other than healthcare for the citizens of the state.

As mandated by federal law, winning bidders *shall not have the authority* to put any restriction on a patient's choice of medical or healthcare provider to encourage patient-doctor relationship, rapport, and continuity of care.

Roles of the Private Sector

The American public owns the United States Universal Healthcare Plan. Like it or not, we, the people, are ethically and morally obligated, and appointed the guardians of our healthcare plan. Each of us must therefore act prudently to protect and make sure its solvency—for ultimately, we are the consumer, provider, and financier of a uniquely American universal healthcare plan.

The public or private sector may be classified either as healthcare consumers, healthcare providers, and healthcare financiers. Eventually, however, at one point in life every American becomes a consumer of medical care.

Healthcare Consumers

The consumers are the American people. It is worth repeating that sooner or later every one of us will be a consumer of the healthcare services in our country. It is inevitable. It is therefore the responsibility of each U.S. citizen to keep the system viable at all times and use its resources wisely. There are, however, among us— those who will be selfish, thinking only of themselves and abuse the system, demanding unreasonable and expensive healthcare services even though not necessary. In those instances, mechanisms to penalize or prevent them from overusing the healthcare system shall be incorporated in the healthcare plan.

Consumers of healthcare services are those seeking preventive, medical, surgical, and dental cares. Preventive healthcare services usually are rendered to children for immunization and the monitoring of children's physical, mental, and emotional developments. Early screening and detection of genetically transmitted diseases and many other childhood diseases are an important part of childhood preventive services. Adult preventive measures are sought by persons with family history of hereditary or familial diseases, i.e., diabetes, heart attack, high blood pressure, stroke, high cholesterol, asthma, and cancer to name a few. It is also directed to people with unhealthy lifestyles, which include the use and abuse of drugs, alcohol, tobacco, overeating, sedentary life, etcetera. Preventive services in these cases are accomplished by regular or periodic physical examinations, blood tests, and other procedures that may be necessary. Another group of individuals who are targeted for preventive care are the healthy adults, age 40 years old and above for women, and 50 years old and above for men, where in addition to the routine physical exam and blood tests, special diagnostic test such as mammogram, flexible proto-sigmoidoscopy or colonoscopy, electrocardiogram (ECG), prostatic specific antigen (PSA), etcetera are recommended. Last but not least are preventive health educational materials and services that should be emphasized and carried by every public health agency and/or private healthcare provider. Health education programs directed specifically to a particular disease entity, e.g. diabetes, high blood pressure, high cholesterol, obesity, etcetera are already in existence. These educational materials are an important part in the general management of the patient's illness and are administered by expert healthcare professionals in the field.

It is a well-established fact that preventive healthcare is the most effective and cheapest healthcare service that should become a part of any healthcare plan. That is why health education of the children from K-12 must be a part of the school curriculum.

Citizens consuming the medical and/or surgical treatments range from the simplest cold to the most complex medical and/or surgical diagnoses. Serious medical, surgical, and dental diseases necessitate the use of advance medical and surgical knowledge, diagnostic equipment, and treatment procedures that are very expensive. Treatment of medical, surgical, and dental diseases, specially the serious life-threatening conditions is where the real healthcare cost is used. Of interest to all is the medical care of the sick elderly person. It poses the doubling or even tripling of medical expenses compared with the younger adults. And with the coming to age of the baby boomers, adequate funding for the healthcare of these new retirees will be difficult, unless the rapidly rising medical cost is aggressively and adequately controlled.

Healthcare Providers

Healthcare providers may be divided into three different groups: the professional, institutional, and healthcare insurance providers.

- The professional healthcare providers are the physicians, osteopaths, dentists, nurses, and other paramedical technicians or staffs, psychologists, social workers, physical therapists, pharmacists, chiropractors, practitioners of alternative and complimentary or integrative medicine, etcetera. They shall be responsible for their own continuing healthcare education and practice *clinical and evidence based healthcare practices* developed from their own fields of expertise. Their fees shall be based on RVU formulated by and for their own professions with the help and supervision of the HSA and DHHS.

- Institutional healthcare providers are the hospitals, large clinics, rehabilitation facilities, nursing homes, hospices; pharmaceutical companies, pharmacies, healthcare instrument manufacturing companies, healthcare supply companies, etcetera shall be encouraged to form networks with the other healthcare industry players to better compete and reduce prices.

- Healthcare insurance providers are the HIC, MCO, HMO, PPO, HPA; insurance consortiums and/or insurance exchanges. They shall also be encouraged to compete with each other and form alliances or networks with each other and with other healthcare industry players for the sole purpose of competing and reducing the prices of healthcare services and products.

Ideally, each type of healthcare provider should accept the *U.S.UHP RVU-Based Fee Schedule* as payment-in-full for services/products incurred by a patient. In the event this kind of reimbursement is not generally accepted, then healthcare providers may be classified as:

- Those accepting the U.S.UHP fee schedule as full payment;

- And those that do not accept the U.S.UHP fee schedule.

Healthcare providers belonging to the second classification shall collect from the patient the total amount of the healthcare services/products rendered. U.S.UHP shall directly pay the patient according to the U.S.UHP fee schedule. The difference between the actual provider bill and the U.S.UHP reimbursement will be the patient's responsibility as it is done these days. In any case, the healthcare provider shall help the patient to secure her/his healthcare benefit from the U.S.UHP.

Healthcare Financiers

The healthcare financier of our healthcare system is the American people. Taxes are paid to the government, which in turn finances its federal healthcare programs. The businesses in most instances subsidize healthcare benefits to employees. Besides federally-funded healthcare programs, businesses contribution to the healthcare system so far is the largest. Last but not least are the private individuals.

Illustrated Implementation

The best method to discuss and show how the proposed Universal Healthcare Plan works is to give an illustration on how implementation shall be conducted. Assumptions, however, have to be made—like in healthcare—the available money is equal to the expenditure, i.e., revenue equals expenditure. The per capita expenditure of $6,280 in the United States was used in the calculation. The assumptions below were based on available figures in 2006.

Assumptions

1. Total U.S. Population: 300,000,000

2. Minnesota Population: 5,000,000

3.   Hennepin County Population:*   750,000

4.   Ramsey County Population:*   500,000

5.   Total HSF Collected: $1,884,000,000,000

*Ramsey County is where the capital city of St. Paul is located. Minneapolis is in Hennepin County.

Based on the above assumptions, the Healthcare Security Agency retains $94,200,000,000 or 5 percent of the total Healthcare Security Fund collected. Ten percent of the HSF5 percent in the amount of $9,420,000,000 will be used for administrative expenses. The remaining $84,780,000,000 will be equally divided as 1) $42,390,000,000 will be used for medical/health education and other healthcare initiatives, and 2) $42,390,000,000 for medical research and development. Unspent HSF5 percent funds shall be deposited in a HSF-RF.

HSF95 percent in the amount of $1,789,800,000,000 will be distributed to the different states of the U.S. by the Health Security Agency based on each state's (per capita) population. Each person in the U.S. will be allocated the amount of $5,966.

Minnesota with five million people will receive the State Healthcare Fund of $29,830,000,000 from the HSA. Five percent of SHF (SHF5 percent) will be retained by the state of Minnesota in the amount of $1,491,500,000. Ten percent of $1,491,500,000 or $149,150,000.00 will be for administrative expenses. The remaining $1,342,350,000 will be equally divided for Minnesota state-sponsored health education and other healthcare initiatives ($671,175,000), and medical research and development ($671,175,000).

The remaining 95 percent of the SHF (SHF95 percent) in the amount of $28,338,500,000 will be equally divided among the five million Minnesotans. This amounts to $5,667.70, is Minnesota's Individual Citizen Healthcare Fund; the allocated amount Minnesota is obligated to use in providing healthcare coverage for each individual in the state.

Hennepin County with 750,000 residents will be entitled from Minnesota SHF agency a budgeted amount of $4,250,775,000. This dollar amount will be available to purchase healthcare coverage for the entire citizen population of Hennepin County of the state of Minnesota.

Bidding Process

Minnesota State will conduct a two-stage bidding process. Prior to the first-step bidding, the state of Minnesota SHF agency will advertise extensively in the state and nationwide to recruit interested healthcare insurers wanting to participate in the bidding process for the County of Hennepin, a few months before the scheduled bidding. To become a legitimate bidder, interested parties have to register with the Minnesota SHF agency. No qualified and legitimated healthcare insurer shall be rejected to participate in the state bidding process.

The state of Minnesota SHF administrator will make a public announcement that the first public bidding for Hennepin County healthcare insurance coverage will be on May 14, 2007 (arbitrary date). Parties interested to bid for the healthcare coverage of Hennepin County will be required to register with the Minnesota SHF agency, and submit their secretly written bid one month prior to the scheduled bidding date to the Minnesota SHF administrator. A registration fee of $10,000 per health insurance bidder will be charged to defray cost.

The following companies for example submitted their secretly written per capita bids and were received by the SHF administrator on or before April 13, 2007.

A.	Allina:	$ 4,000
B.	Blue Cross Blue Shield of MN:	$ 3,900
C.	HealthPartners, Inc.	$ 3,800
D.	Kaiser Permanente:	$ 3,700
E.	Medica:	$ 3,600
F.	Preferred One:	$ 3,500
G.	Prudential:	$ 3,400
H.	U-Care:	$ 3,300

| I. | United Healthcare: | $ 3,200 |
| J. | Etcetera Health Plans: | $ 3,100 |

Etcetera Health Plans submitted the lowest bid at $3,100.

On May 07, 2007 Minnesota SHF administrator notifies all the registered bidders and makes a public announcement that the lowest written bid is $3,100 without mentioning the name of the lowest bidder and the date of the second bidding process publicized.

An open forum public bidding (reverse auction) was held at the Hennepin County Government Center on May 14, 2007. The Minnesota SHF administrator (who chaired the bidding process), the Hennepin County Secretary, the registered bidders and interested members of the Hennepin County public, attended it. The rules of the bidding process were stated publicly and bidding started. HealthPartners offered the lowest public bid at $2,900 per capita or $2,175,000,000 for the entire Hennepin County Population of 750,000 residents. The savings for the Minnesota SHF agency for Hennepin County Healthcare alone is $2,075,775,000.

A similar bidding process was conducted at Ramsey County. This time the winner of the bidding process was Kaiser Permanente.

SHF administrator then schedules another public bidding for another county—at the county's government center—following exactly the same procedure as that of Hennepin and Ramsey counties. HealthPartners and Kaiser Permanente formed an alliance to have better business leverage in negotiating for cheaper prices for drugs and other nationally-distributed health-related products from manufacturers such as prescription drugs. Each insurer also used bidding with the other healthcare providers, for better prices of their services or products—a strong incentive for the insurer to make, to increase profitability and for the provider to compete, to get the business, which is substantial.

Citizen Healthcare Registration

Immediately after the second-bidding process is finished, a winner is officially declared. Hennepin County is notified by the Minnesota

SHF agency to start enrolling all the county residents for the United States of America Universal Healthcare Plan using the federally designed and approved U.S.HPA application form. Registration shall be conducted at all the area Hennepin County Service Centers. Registration fees shall be collected to cover cost. Initially this may take a several months to complete registration of the residents of Hennepin County.

Upon completion of the enrollment, Hennepin County will send a copy of all the registrants in Hennepin County with their NHID numbers/healthcare account numbers to HealthPartners, Inc. Each Hennepin County resident registered with the United States of America Universal Healthcare Plan shall be issued the U.S.UHP ID card by Hennepin County. As soon as Hennepin County has issued the U.S.UHP ID card to all the citizens, Hennepin County notifies the Minnesota SHF agency, which then pays $2,175,000,000 to HealthPartners. HealthPartners, Inc. from then on will be responsible for paying medical bills incurred by all Hennepin County residents; just like any insurance or managed-care organization would handle them today.

Winning Bidder's Responsibilities

HealthPartners shall not impose any restriction as to which primary or specialty healthcare provider a patient wants to see to establish a patient-doctor relationship and continuity of care. Nor shall HealthPartners interfere with the patient-doctor relationship and/or write policies that will hinder the healthcare provider's day-to-day care of their patients.

HealthPartners and Kaiser Permanente shall do their best to encourage and negotiate with healthcare providers to accept U.S.UHP RVU-based fee schedule reimbursement formula. It shall make sure at all times that healthcare expenses are affordable—by forming alliances, consortium, and/or exchanges with other insurers (winning bidders) in the state of Minnesota or even in the U.S. to develop drug formulary, and networks with other healthcare providers, e.g., large groups of diagnostic or procedure-oriented clinics, pharmacies/drug companies, hospitals, physical therapy and chemical dependency rehab centers, mental health facilities, chiropractic and complementary, alternative or integrative medical clinics, medical supply companies, home care companies, nursing home/hospice facilities, long-term care services, etcetera—and negotiate for the lowest possible prices and fees to insure company profits.

HealthPartners shall honor all out of county, state, or country healthcare service(s) rendered to any Hennepin County resident and shall be responsible for paying the healthcare bill of the resident—minus required co-payments. HealthPartners may also introduce other healthcare initiatives such as preventive healthcare and lifestyle management to improve long-term profitability.

The contract of HealthPartners, Inc. with the Hennepin County and/or with the Minnesota SHF agency shall be for a period of three to five years, or as determined later as appropriate. During the period of the contract with Hennepin County, HealthPartners, Inc. shall be required to have a reinsurance policy, and submit to the Minnesota SHF agency a copy of the reinsurance policy that is in effect for the entire length of the contract.

Healthcare Provider's Responsibility

Co-payments will be charged and paid by the healthcare consumer at point-of-service for every healthcare provider's visit and prescription drugs. The co-payment for primary care visit will be X dollars and for specialty visit or consultation (with the exception of emergencies) without a referral from a primary care provider, Urgent Care or Emergency Room visits will be three to five times more to discourage over-utilization of these services. The co-payment fee will be collected and remitted to the Healthcare Security Fund.

Healthcare Reserve Fund

The Healthcare Reserve Fund for Minnesota established by the state SHF agency as mandated by the federal healthcare law enacted shall be equal to the calculated amount necessary to operate and finance the Minnesota SHF for ten years. For

Minnesota that will be $280,338,500,000. Amount of the reserve shall be adjusted annually at the rate of inflation. The bidder's registration fees, savings from the bidding process, and interest from investments shall fund the state SHRF. Excesses in the projected 10 year state SHRF shall be used for other healthcare programs or initiatives that are deemed necessary.

Co-payments made by patients for healthcare provider's office visits, and pharmacy providers shall collect prescription drugs and remitted to the Healthcare Security Fund. Co-pays shall be predetermined but could be adjusted or even eliminated by the providers when dealing with poor citizens of state based on income and ability to pay.

Registration fees generate from healthcare insurances participating in the state's bidding process and the amount of money saved shall be used as part of financing the state SHRF. Other sources of money for the SHRF shall be the unspent budgets from administrative, educational, research, and development.

SHRF investments shall be conducted by the SHF committee with the help of a financial employee (chief financial officer) and/or hired investment advisers.

Optional Healthcare Plans for the States

Some healthcare policy experts like Stuart Butler, PhD, and David Kendall favor incremental changes in healthcare reforms through the states to accomplish universal healthcare. They see Americans and particularly the U.S. Congress as highly resistant to drastic changes. Massachusetts is frequently cited as an example of a state that has taken the initiative to reform the healthcare system and provide universal healthcare coverage for the citizens of Massachusetts. Whether Massachusetts is going to be successful remains to be seen.

Failure on the part of the federal government to legislate laws establishing a U.S. Universal Healthcare Plan should not discourage individual states from forming, developing, and enacting their own State Healthcare Plan based on the framework of a universal healthcare plan hereby proposed. It will be much better, however, if states can band together and create Regional

Healthcare Systems. Example is for Minnesota, Wisconsin, Iowa, North Dakota, and South Dakota to form the Upper Midwest Regional Healthcare Plan. Another option is for a state to form consortium of as many states willing to participate in such a healthcare plan.

Funding will require the state government to request Medicare and Medicaid to enroll each senior resident and other Medicare and Medicaid recipients of the state, and transfer the allocated funds directly to the state healthcare plan, similar to how private insurances make arrangements with Medicare and Medicaid to cover state senior citizens and welfare patients in their healthcare plans. Employers, employees, and self-employed contributions shall be collected by the state as State Healthcare Fund Tax (SHFT). The co-pays collected by providers, and including the state government's contribution shall also be part of the SHF.

For humanitarian reasons some state may want to provide healthcare coverage to illegal immigrants. Such states may do so by establishing a separate state-sponsored healthcare system funded entirely by the state. At no time shall funds from the SHF and/or SHRF be used for such/other purposes.

Proposed Alternative Healthcare for States

I.   State government, through the act of the State Congress creates into law a State Healthcare Superfund (SHS) to pool all the financial resources assigned for healthcare—the state's share of the Medicare and Medicaid programs distributed periodically from the federal government; the state health expenses budgeted for the state indigent population, and from all employers, employees, and taxpayers of the state.

II. State government establishes a *State Healthcare Agency (SHA)* that shall administer the State Healthcare Laws (SHL).

- The SHA shall have an administrator (CEO) appointed by the state governor and confirmed by the legislative branch of the state government; 12 committee members (directors) appointed by the governor and confirmed by the legislative branch of the state government.

- The SHA shall retain 5 percent of the total SHS fund for administrative expenses, health education, and medical research and development.

- It shall develop and formulate *State Healthcare Policies (SHP)* [with the cooperation of the State Health Department] written as general principles rather than as specifics, and avoiding details whenever possible. Such SHP shall be presented and reviewed for approval or disapproval by the state Congress. If approved by the state Congress and signed by the State Governor, the SHP becomes the SHL.

- It shall supervise and impose the SHL, especially the two-step bidding process in all healthcare business transactions at all levels of the healthcare system, and other laws directed to current or future problems that threaten the viability of the state healthcare programs.

- Savings or unspent funds shall be use to fund the State Healthcare Reserve Fund (SHRF).

III. Implementation:

- The SHA after removing 5 percent from the total SHS fund shall use the remaining 95 percent to purchase healthcare insurance for all the state resident citizens.

- SHA conducts a competitive two-step bidding process for a designated state, city or county population to buy healthcare coverage for the specified citizen population from private health insurances, HMOs, PPOs, etcetera.

- The winner in the bidding competition shall conduct the two-step bidding in purchasing healthcare provider services and products.

- Winning company shall provide healthcare coverage for all state or county citizens included in the bidding process.

- County registers and issues a State Healthcare ID card for every citizen resident in the area to easily access healthcare services anywhere in the U.S.

- State citizens shall have the freedom to choose their healthcare provider, and access the healthcare system by presenting their healthcare ID care to the provider.

- SHA shall fund the health education curriculum from elementary to high school education; and medical research and development.

- All saved or unspent monies shall be used to fund the State Healthcare Reserve Fund (SHRF); accumulating all unspent healthcare funds to satisfy a ten year projected healthcare budget for the state.

- Winning healthcare insurance company shall be compelled to provide a reinsurance policy from big dependable reinsurance companies.

National Healthcare ID

The federal government shall have a *N*ational *H*ealthcare *I*dentification *(NHID)* system. It must be uniformly applied to all the states of the United States of America. It shall have characteristics that could easily identify an individual as to where such a person comes from and what healthcare provider or health insurance company coverage she/he has.

When swiped in a healthcare provider's point-of-service or office the computer screen at the front office will show an exact replica of the patient's NHID with her/his picture, fingerprint, signature and other biometrics that are included. The front office then enter the point-of-service/office code and the patient is ready to be seen by his/her provider

A system of identifying that will satisfy the above requirements had already been developed and a variation of the system is herein presented.

In the 1970s there were 146 recognized independent nations of the world. Assigning a number to each country according to the alphabetical order of their names will give us the following possible codes for each country: Canada is # 020; Great Britain # 049; Japan # 067; Mexico # 087; Norway # 100; Philippines # 105 and the U.S. # 136. Any newly formed countries will be assigned a number at the time of their official recognition as independent nations. These future nations, X, Y, and Z will be numbered as 147, 148, and 149 respectively. As of 2012 there are 196 known countries in the world; 193 are members of the United Nations. As stated, the number of countries changes. The system of coding is flexible enough to accommodate new countries as they emerge.

Another method is to assign a number or code to a country when such countries begin adopting a standard universal system of identification. As an example, if the U.S. adopts first the system followed by Canada and then by Mexico, the code numbers for these countries will be:

U.S.:	**001**
Canada:	**002**
Mexico:	**003**

The proposed National Healthcare Identification—based on the original universal system of identifying, is as follows.

000 – 00 – 00 – 00000000 – 0000 – 0000000
A   B C D   E   F

The word state also means province, territory, or protectorate of a country. The word county also refers to and includes a district or its equivalent and a U.S. Embassy or a U.S. Consulate Office in a foreign country.

A is the code representing the country of residence and point of registration of an American citizen. *B* is the state of the country of residence and place of registration. *C* is the county of the state or a U.S. embassy/consulate office in the country of residence and registration of an American. *D* is the date of registration in the county of the state point of registration. *E* is the number of registrants in the county of the state at the time of registration per day/twenty-four hours. *F* is a modifier representing the registrant's current/latest country, state and county of residence. *F* when arbitrarily separated will be represented as—000—00—00. The first three digits (– 000 –) is the code for the country, the second group (– 00 –) is the code of the state and the last two digits (– 00) is the code given to a county of a state in the U.S. or a U.S. embassy or consulate office in the country where an American healthcare registrant resides in a foreign country. In cases where the registrant stays in the same county, state and country of registration the *A— B—C* and *F* are the same. In persons who change residency the *F* will change as frequently as they are relocated.

The four digits number in the *E* group may be too many or too few. It can be adjusted to the appropriate number of digits depending on the populations of the counties. The number of digits in this group once determined shall be applied uniformly for all the NHID. If the number of digits in *E* group is three (000), this will give up to 999 registrants in one location per twenty-four hours. And if it is four (0000) digits, this will allow up to 9999 registrants per twenty-four hours in one place of registration. What is the desired number of digits in *E* has to be determined.

To illustrate, let's take as an example the countries of United States of America and Canada and assign the following codes:

U.S.:   **001**

Canada: **002**

U.S. has fifty (50) states and Canada has twelve (12) provinces and territories. The codes assigned to the different U.S. and Canadian states were based on the dates these states became recognized states of the United States of America or Federation of Canada. U.S.: Delaware is the first (1st) state of the Union and Minnesota is the thirty-second (32nd). Canada: Alberta is the ninth (9th), British Columbia the seventh (7th) and Ontario is the first (1st). The code numbers for these states shall be:

U.S.:   **Delaware**   **01**

                  Minnesota   **32**

Canada:	Alberta	09
	British Columbia	07
	Ontario	01

Delaware has three and Minnesota 86 counties. The codes given to the counties were based on the alphabetical orders of the names of the counties.

Delaware:	Kent	01
	New Castle	02
	Sussex	03
Minnesota:	Hennepin	27
	Ramsey	62

There are several U.S. embassy and consulate offices in Canada. As an example let's arbitrarily assign codes to these embassy and consulate offices in the different provinces.

Ontario:	Ottawa	01
	Toronto	02
Alberta:	Calgary	03
B. Columbia:	Vancouver	04

To illustrate let's take for example a few Americans residing and registering in the different U.S. and Canadian states. Canada can easily be any country in the world.

- Alexander Nelson resides in and is registered for his National Healthcare ID in Hennepin County, Minnesota, on January 01, 2007. He is the first registrant in Hennepin County, Minnesota. His NHID is:

001 – 32 – 27 – 01012007 – 0001 – 0013227

Alexander continued to live in Hennepin, Minnesota, and never moved out of the area. His NHID never changed.

- James Taylor resides in and registered for his NHID in Hennepin, Minnesota, on July 10, 2007. He is the 1234th registrant that day in Hennepin, Minnesota. His NHID is:

  001 -32 – 27 – 07102007 – 1234 - 0013227

  James later moved to Ramsey County, Minnesota, two years later and never moved again. His new NHID is:

  001 – 32 – 27 – 07102007 – 1234 - 0013262

- Abigail Ford resides in and registered for her NHID in New Castle County, Delaware on February 12, 2007 and is the 234th registrant in New Castle, Delaware that day. Her NHID is:

  001 – 01 – 02 – 02122007 – 0234 – 0010102

  She later moved to Ottawa, Ontario, Canada nineteen years later. Her new NHID is:

  001 – 01 – 02 – 02122007 – 0234 – 0020101

  Abigail ten years later moved back to the U.S. in Ramsey County, Minnesota. Her new NHID is:

  001 – 01 – 02 – 02122007 – 0234 – 0013262

- Isabelle O'Brian, an American citizen resides in and registered in Vancouver, British Columbia, Canada on November 28, 2008 and is the fifteenth registrant that day. Her NHID is;

  002 – 07 – 04 – 11282008 – 0015 – 0020704

  A year later she moved to Calgary, Canada. Her new NHID is:

  002 – 07 – 04 – 11282008 – 0015 – 0020903

Eight years later she was transferred to Toronto, Ontario, Canada. Her new NHID is:

002 – 07 – 04 – 11282008 – 0015 – 0020102

Fifteen years later Isabelle moved back to the U.S. in Sussex, Delaware. Her new NHID is:

002 – 07 – 04 – 11282008 – 0015 – 0010103

The modifier *F* will effectively track the current residence of any American NHID holdeR&Determines what or which insurance company is responsible for medical bills incurred by any American NHID holder. The *A – B – C – D – E* groups become the permanent medical record of the registrant. American citizens registering in foreign countries shall have their medical records kept in a computer database in the U.S. under the auspices of ALAHA.

Note that the person's NHID numbers/healthcare account number does not represent any personal information. The NHID A – B – C – D – E groups of numbers are meant as a systematic method for filing and accessing medical record and any other information related in the most effective execution of the healthcare system. The – *F* – number on the other hand is a useful and practical means of tracking down the participant's location/movements and health insurance coverage. An issued NHID card then becomes an approach of quickly and efficiently executing a healthcare plan that is as big as the U.S. with a population as mobile as we are.

The NHID card shall have the following features.

- All NHID cards shall be uniform in general content and design. NHID cards shall be a picture ID technically tamper-free or counterfeit-proof and difficult to copy or imitate. It shall have the appearance and size of a credit card.

- The NHID card shall be best known as the *U.S.UHP ID CARD.* It shall be made of durable heavy duty (plastic) or equivalent material that should last at least five to ten years.

- The front face of the ID card shall feature the person's facial picture with the top inscribed the U.S.UHP ID CARD, her/his name and the ID number in numeral and/or bar codes strategically located. The back shall contain a magnetic strip or its equivalent containing personal data necessary only for billing and record keeping purposes similar in today's credit cards. No detailed personal data as taken in the U.S.HPA form shall be included in the NHID card. It shall contain just enough information for easy access of patients, and for efficient billing and medical record keeping applications that shall prevent the use of paperwork and allow easier electronic transactions.

- To increase the authenticity of the NHID card other biometrics maybe included such as fingerprints, signatures and so on as other means of identification.

- The NHID card shall be issued by the states only to all U.S. citizens verified by birth certificates or naturalization papers and aliens with permanent resident visa or refugee status verified by their immigration papers and personal interview(s) if judged necessary. Other legal aliens such as tourists, foreign students, business persons, etcetera coming to the U.S. shall not be eligible for issuance of the NHID card.

- The counties of each state shall be the issuing agents for the NHID card of the state using the U.S. Healthcare Plan Application form as a standard method for collecting individual personal data and the basis for issuing an individual NHID card.

- The name of the county and state must appear in the U.S. Healthcare Plan Application form to easily identify and authenticate where the registration was done.

- Registration of an individual in more than one county and/or other government agency shall not be allowed and will be strictly prohibited. Doing so will result in fines/penalties and be treated as a misdemeanor or felony.

- The NHID number becomes the Medical Account Number and/or Medical Record Number of individual Americans. This will facilitate the transfer of medical record/information, electronic or otherwise, upon properly authorized and executed request of the individual and/or her/his healthcare provider.

- There shall be provisions in the law to outlaw with stiff penalties the unauthorized transfer and/or use or misused of individual medical record and/or information.

- Counties of states receiving transferring residents of other counties from the same or other states shall verify their new residents' identity from the county from which they were originally registered or came.

- Americans transferring to other counties, whether in the same state or of other states, shall report and renew her/his NHID card to the new home state's county one month before their birth date to prevent or avoid lapses in her/his healthcare coverage.

- Every American age 21 years and above shall be required to re-register for her/his U.S.UHP benefit every five years to update her/his personal data base. Children shall re-register every two years; because of the rapid changes in the children's physical features.

III

Evolution of America's Healthcare Crisis

Historical Background

Post World War II healthcare in the United States was primarily financed privately by individuals. Patients paid their own medical bills. Those with private health insurances bought the policy themselves. Few companies provided medical care benefits to employees. The federal government played a limited role in the public medical care system; the most was providing medical care to veterans and federal employees through the Veterans Hospital System and federally-funded health benefits for employees. There were public clinics and hospitals but not many to play a considerable effect in the healthcare economics of the day.

Medical care, whether in public or private outpatient clinics or hospitals practices, was administered by medical professionals with patient care as the single driving force in the practice on medicine. The American public on the other hand exercised good judgments in the use of their medical resources. They listened carefully and took their doctor's recommendations seriously and without questions, used common sense on how to take care of minor ailment themselves, and never demanded the latest and most expensive test, procedures, or treatments of the time. It was not unusual for doctors to visit and treat their patients at home, and had the time to talk and listen to them. The economics of medicine was farthest in their doctor's mind. Medicine was then a profession.

Professional satisfaction among physicians was very high. It was the "Art and Science of Medicine" at its peak.

Patients who were unable to pay for their medical care, for whatever reason, were nonetheless treated. One's ability to pay was not an issue to receive medical care. Physicians saw their indigent patients and treated them with respect and dignity. Sample medications from pharmaceutical companies were generously given to those who could not afford their medicine. Hospitals allocated charity beds indistinguishable from regular-paying hospital beds. Most, if not all, who needed treatment received outpatient and/or in-hospital medical care. Medical care was reasonably priced and affordable. The average hospital cost per day in the early 1960s was below $100 and $30 additional for ICU.

In 1965, Medicare and Medicaid were enacted. The recipients of the new federal programs began using the benefit they newly acquired in large numbers. Medical expenditure started to swell. A Minnesota physician, Dr. Paul Ellwood, who was the executive director of the American Rehabilitation Foundation and the Sister Kenny Institute of Minneapolis, saw the passage to the Medicare and Medicaid as the beginning of socialized medicine and was deeply concerned about the quality and practice of medicine in the coming years. As early as in 1960 Dr. Ellwood was already thinking of "health services" managed and administered by professionals in the medical fields. The concept of health maintenance and other managed-care organizations was later born and gained popularity. The *Health Maintenance Act* of 1973 was enacted by the Minnesota Legislature. The steady climb in medical care costs also caught the attention of the federal government which began looking into the effects of *Health Maintenance Organizations (HMOs)* and other Managed-care *Plans*. Dr. Ellwood, considered the father of HMO in the early 1970s, initiated talks with the Department of Health and Human Services. This culminated in the legislation of the federal Health Maintenance Organization Act of 1973. Grants were given to study and financially support these organizations. The HMOs were unregulated and became, financially, very successful. The insurance companies saw the opportunities and developed their own HMOs. By 1997 about 75 percent of HMOs were for-profit managed-care companies up 18 percent from the

1980s. This staggering statistic was thought by some as representative of one of the largest transfer of public funds to the private sector in U.S. history.

Large employers began providing medical care subsidies as part of their employee's contracts because of cheap health insurance premiums. It rapidly became a standard employee's benefit package in the early1970s and flourished through the mid-1990s. Both the government and employer's healthcare programs unleashed the insatiable public appetite for medical care as individuals became less and less personally involved in the cost of their healthcare. People began abusing their government entitlements and employee's medical benefits. They frequently visited their doctors for the slightest medical conditions such as colds, insect bites, skin abrasions and bruises, etcetera, used the ambulance for minor cuts or accidents and the emergency rooms for non-emergent illnesses; not as a matter of necessity but rather for the patient's or relative's conveniences. Request for the latest medical diagnostics and treatments accelerated as patients increasingly become isolated and/or ignorant of the actual cost of their medical care.

To make the healthcare industry more efficient in meeting public demands, health administrators and business managements did what they knew best to make the industry grow. Medical facilities performing below business expectations were closed, and consolidation of hospitals and clinics began occurring in the 1970s through the 2000s. This was followed later by merging of health insurances, HMOs, and other forms of managed-care organizations, pharmaceutical, medical manufacturing companies, and other health-related companies resulting in fewer but much larger, powerful, and politically active health institutions.

With increasing usage for healthcare services in the 1970s, physicians started forming bigger and larger medical practices as either single or multi-specialty groups. Clinics and hospitals expanded their facilities through acquisitions of smaller clinics and hospitals, and acquiring the latest expensive medical equipment only to be discarded in a few years because of newer and more technologically-advanced medical instruments. There was a buying and leasing frenzy of medical office spaces and medical apparatus by big medical groups, clinics, and hospitals. Large clinics and hospitals that had the financial muscle constructed big medical complexes either for doctor's offices or clinics adjacent to hospitals. Old hospitals were renovated to meet the new standards. In some instances, old hospitals were demolished and modern up-to-date hospitals went up to replace the old, and to satisfy growing needs. Expensive medical educational and follow up programs—necessary or not—for patient cares were developed and aggressively marketed. The common reasoning behind all the pricey business decisions made by healthcare managements was to stay competitive; a rationale resulting in increased medical costs.

HMOs' administrative expenditures rose to as high as 30 percent. In spite of that, the common thinking was: "There is lots of money to be made." Profits were plowed back into these costly business activities instead of lowering costs or sharing them with the consumers through reduced insurance premiums, for example. This form of competition was premised on the availability of inexhaustible healthcare resources. It should have been based on a realization that healthcare resources are finite—that no economy can sustain steady [at times double digit] increases in healthcare costs without breaking the system.

Even when the economy slowed down and went into recession, managed-care institutions and the rest of the healthcare industry reported unprecedented profits. In later years, managed-care became and is still one of the driving forces in the ever-accelerating healthcare expenditure in the United States.

On the other hand, there came a severe squeeze in available medical office spaces elsewhere in the communities. Buildings designated as medical offices, the rents/leases often were priced 40 percent to 60 percent more than ordinary offices for other businesses. Doctors' overhead expenses increased alarmingly in all the areas of medical practice. They had no other recourse but to raise fees and passed the overhead increases to their patients.

The role of HMOs and other forms of managed-care organizations, as players in the healthcare industry, in the beginning were insignificant. As they established their "roots" in the healthcare system they became a formidable force to contend with. Initially they effectively controlled their medical expenditures as a group. Most healthcare experts thought they found the answer to the nation's ballooning medical care cost. In recent years, however, the HMOs and managed-care healthcare providers are behaving like their predecessors in the healthcare industry. They are operated as big businesses with the primary goal of making a profit. They are now considered by many as part of the big problem.

Going back in the early1960s, hospital interns and resident physicians were paid a minimal salary called a stipend. Work demanded from 88 to 112 hours per week schedule depending of the number of house staff (interns and residents) in the training hospital. Stipends were from $100 to $ 300 a month, not enough to support oneself. In the late 1960s, things changed. Hospital house staff activism around the country for better working conditions and pay commenced. Stipend went up in the late 1970s and steadily increased to the present levels of around $30,000 to $60,000 annually. The young physicians' demands were not isolated cases. Soon the nurses and other paramedical personnel who were underpaid compared to employees from other businesses with the same education and work experiences, followed the examples of the interns and residents. There were significant increases in salaries, and yet even to date these healthcare professionals are still paid much less than their counter parts in the business world. Although these stipend and salary increases—well deserved by these healthcare professionals and still not in par with other business professionals—no doubt are contributing to the steady financial burden of the healthcare system.

Like any other organization, the American healthcare system has pros and cons. With all its faults, our healthcare has one big redeeming quality. It can deliver in most every instance much needed medical care; whatever kind, whenever indicated, and wherever in the U.S.—without delays.

America's Healthcare Statistics

The reader is forewarned that she/he will encounter differences in the values of the statistical data cited. The differences are because the information came from various sources—surveys, projections, and estimates using variety of techniques and parameters of measurements, and were done in different years.

Study after study, year after year, show skyrocketing increases in healthcare cost in our country. In 1998 our country spent $1,016,000,000,000 with an average per capita healthcare expenditure of $3,759. There was a wide variation from state to state with District of Columbia leading the nation at $6,656 followed by MA at $4,810, NY at $4,706 and Utah the least with $2,731. Minnesota's per capita healthcare expenditure then was $3,966, or the 12th most expensive state. (Ref: Kaiser Family Foundation State Health Facts Online)

The Center for Medicare and Medicaid Services reported that healthcare spending increased 13 percent in the last decade and jumped to 14 percent in 2001. (Ref: WAIS Document Budget of the U.S Government)

Each American by the year 2002 paid an average of $5,440 for healthcare with a total of $1,600,000,000,000 for the entire country. (Ref: Boston Herald Business.com)

In 2002 for small businesses in the U.S., an average of 9.4 percent of their annual revenue was spent on healthcare for their employees. (Ref: SBA)

GM alone in 2002 reported a projected $67,500,000,000 obligation for retiree and employee healthcare cost. GM spokesman Jerry Dubrowski said, "Before a car leaves our factory it's got a $1,400 cost disadvantage relative to an overseas model." (Ref: Boston.com/Business)

From the year 2000 to 2003, healthcare insurance premium increased by more than 50 percent averaging about 17 percent per annum. (Ref: CNN Money)

In 2003, 45 percent of the nation's healthcare dollar came from public and 55 percent from private funds. Specifically these were 17 percent from Medicare; 16 percent from Medicaid; 12 percent from public funds for Workman's Compensation, Public Health Activity, Department of Defense, Department of Veterans Affairs, Indian Health Services, and State and Local Hospital and School Health services; 36 percent from private insurances; 14 percent from out-of-pocket individual expenses, and 5 percent from other private health funds including industrial, construction, and philanthropic revenues. (Ref: Congressional Budget Office)

By 2003 company sponsored health insurances contributed an average of $508 for single and $2,412 for family coverage. Adding the employee's contributions to their own healthcare expenses, the total is $3,383 for single and $9,068 for family healthcare insurance coverage. (Ref: Business Finance Magazine.com)

From 2003 to 2004 Medicare increased an estimated 8.2 percent and Medicaid by 10.5 percent; an average of 9.35 percent for these two federally funded healthcare programs. Estimated expenditure in 2004 for Medicare is $265,900,000,000 and Medicaid at $182,300,000,000 totaling $448,200,000,000. (Ref: WAIS Document Budget of the U.S. Government) (This does not include the recently approved Medicare Drug Benefit of $400,000,000,000 in the next ten years.)

A closer examination how fast Medicare and Medicare have grown over the years and how much it will continue to grow in the next decade, the following estimated figures from the Congressional Budget Office expressed in billions of dollars should tell the story.

Year	Medicare	Medicaid	Total
1965	000.00	000.30	000.30
1975	014.10	006.80	020.90
1985	069.70	022.70	092.40
1995	177.10	089.10	266.20
2004	297.40	176.20	473.60
2005	332.00	184.00	516.00
2006	385.00	192.00	577.00
2007	437.00	203.00	640.00
2008	462.00	221.00	683.00
2009	491.00	239.00	730.00
2010	527.00	260.00	787.00
2015	785.00	387.00	1,172.00

Assume conservatively that the average increase in the national healthcare expenditure from 1998 to 2004 is 9.35 percent—the projected total healthcare disbursement as a nation for 2004 is $1,780,000,000,000 or $6,573 per capita. Even at a conservative 6 percent annual rise in healthcare cost from 2004 to 2010 the cost will be a staggering amount of $2,550,000,000,000 or $8,500.00 per capita. (Based on a U.S. population of 300,000,000 by 2010)

Another way of looking at the dramatic increases in healthcare cost in the United States is by examining the **Gross Domestic Product (GDP)**. In 2000, the healthcare expenditures in relation to the GDP was 13.1 percent; 2001, 13.8 percent; 2002, 14.6 percent and 2003, 15 percent. Comparing the healthcare expenditures of the different countries in relation to GDP in 2003, U.S. again led the way at 15 percent, followed by Germany at 11.1 percent; Switzerland at 10.9 percent; France, 10.5 percent; Canada, 9.9 percent; Sweden, 8.4 percent; Japan, 7.6 percent; and England at 7 percent. All these countries are already experiencing rising healthcare expenses, and with anticipation of large increases as the population grows older. The United States by far spends the highest for healthcare, followed in not so closely second place by Germany.

Some figures mentioned in a recent medical seminar at the University of Minnesota estimated that the U.S. healthcare will cost $5,000,000,000,000 the year 2015. That's $16,129.00 per person in the United States assuming a U.S. population of 310,000,000 people. Can the average American afford this amount?

A frequently quoted 46,000,000 Americans or 15 percent of our population are uninsured. Many are young middle class between the ages of eighteen to twenty-three years old. Most are recent high school and college graduates who are unemployed.

Lastly, it was recently reported that Medicare will be financially bankrupt/insolvent by the year 2019.

The Sources of Healthcare Expenditures in the U.S. are: (Ref: The Medical Expenditure Panel Survey)

Private Insurance:	44.5%
Medicare:	21.1%
Out of Pocket:	18.0%

Medicaid:	8.6%
Others:	7.8%

Here's how America's healthcare dollars are spent. Again there is a substantial spread how the money is spent from state to state. (Ref: Kaiser Family Foundation)

Hospital Care:	37.4%
Health-Care Professional Services:	29.1%
Drugs & Medical Non-durables:	12.1%
Nursing Home Care:	8.6%
Dental Services:	5.3%
Home Healthcare:	2.9%
Medical Durables:	1.5%
Other Personal Healthcare:	3.1%

With a lagging economy, outsourcing of good paying jobs to foreign countries, high unemployment and huge trade deficits, just to name a few, will the American people be able to support and/or maintain these large increases in their healthcare?

These are frightening statistical realities that will significantly impact on the nation's well-being in personal health and economic stability. American companies will be at a great disadvantage competing in the world market because of the tremendous monetary burden healthcare has on its industries.

Unless the factors causing the increases are firmly reined in, appropriately dealt with and resolved *as soon as possible*, America's healthcare system will break down under enormous financial pressure in the coming years. The question is: "Do the U.S. Congress and the President of the United States have the political will, courage and strength to make the change?"

Bleak as the healthcare system may seem, there are silver linings in the horizon. *Many smaller managed-care organizations had reported in the medical literature that with tight management of financial resources in the presence of a "nice mix" of clients in* *sufficiently large number, the cost to healthcare can be lowered to $1,800 to $2,000 per capita per year.* Americans are now taking healthcare issues seriously and are openly discussing how the system can be rehabilitated.

Here is a comparative healthcare costs per capita spending here and abroad in 2001 and 2004:

Country	2001	2003	2004
U. S. A.	$ 4,540.00	$ 4,887.00	$ 6,280.00
Germany	$ 2,677.00	$ 3,817.00	$ 4,387.00
Switzerland	$ 3,248.00	$ 3,781.00	$ 4,208.00
Sweden	$ 2,800.00	$ 3,000.00	$ 3,300.00
France	$ 2,561.00	$ 3,300.00	$ 3,776.00
Japan	$ 1,984.00	$ 2,724.00	$ 3,768.00
Canada	$ 1,837.00	$ 2,792.00	$ 3,572.00
Britain	$ 1,457.00	$ 1,992.00	$ 2,357.00

At one Minnesota HMO for the year 2003, each enrollee's personal healthcare expenditure was around $3,400. Another Minnesota HMO in 2005—with less than 200,000 members—offered Medicare supplement with comparable coverage at about $1,300 annual premium. Other HMOs charge from $3,000 to $4,200 annually for similar products offered by the first.

In 2004 the total healthcare expenditures in the U.S. were $1,900,000,000,000 or per capita health expenditure of $ 6,280.

Problems and Solutions

Ideally, any national healthcare plan that encompasses all the citizens of a country, in order for it to last for generations, must have features that would resist and adapt to unforeseen financial demands it may later encounter as it moves into the future. To achieve such a healthcare system is to incorporate safeguards based on careful examinations of the different factors maligning our present healthcare. Because the problems are too complicated and often interrelated, public participation in a national exchange of ideas to review and resolve all present and future identifiable troublesome aspects of the healthcare system has to occur—insuring extensive analysis of the problems from different points of view offered by the American people. The proposed universal healthcare plan attempts to address healthcare problems deemed as major factors responsible for the escalating cost of medical care in this country.

Healthcare Abuses by Patients

There are significant numbers of patients who are high users of healthcare services. They may be classified into four different groups. The first are patients who have no responsibility paying any amount for their healthcare. Second are those who demand the newest and the most expensive diagnostic tests, treatments, and drugs. Third are the elderly who can not accept that growing old is a natural phenomenon, and as a result, they demand everything be done without regard to expenses in the face of multi-organ failure and imminent death. Lastly are the relatives who request only the best and most intensive care for their love ones regardless as to the benefit or cost of the treatments.

People who have no financial stake in their healthcare often visit their providers with the least minor complaint they have. They run to their doctors every chance they can. Some visit their doctors so frequently they may average one visit every week or two, year round. They think and believe it is their birthright to be treated as frequently as they please with the best of possible care—without realizing—at the expense of their fellow citizens. When confronted with their frequent visits, they argue they're entitled because the government is paying for their healthcare, and it is the law.

Individuals, such as these, must be given some financial responsibility for their personal healthcare. As their fair share to the system, a predetermined token fee for their care, collected as co-pay at point-of-visit will discourage over utilization and help to reduce the unnecessary and frivolous overuse of healthcare resources.

There are also a significant number of patients who demand only the best and most expensive care. They tell their doctors exactly what they want done. They are usually well informed about their symptoms, have self-diagnosis, and think they know better than their physician. They argue they're paying a lot for their medical care and therefore are entitled to the best and most expensive tests and treatments available in medical management. They do not realize that their demands far exceed the premium cost of their health insurance and out-of-pocket expenses annually. They are very intimidating, especially in this age of "the customer or patient is always right," and litigations, adding more stress in an already very stressful work environment for providers.

The elderly patients who have not accepted the fact they are old and their bodies are failing because of their age form another group of demanding patients. They expect every failing part of their body be aggressively treated or repaired. It is not uncommon to have patients in their seventies, eighties, or even nineties who have undergone two to four coronary heart by-pass or other revascularization procedures, hip and/or knee replacements, etcetera. In most instances these repeated major medical and surgical treatments are not beneficial to the patients, and are not cost effective. As an example, 5 percent of Medicare recipients use 60 percent of Medicare resources for the treatment of diseases of the old age that can not be won. Most die within three to four months—a real waste of medical resources that could be better used for the young, for prevention, for education and research.

Lastly are the patient's relatives who want every possible means done to save the patient without any consideration to appropriateness, benefit, or cost of the treatments. It would appear that their only aim is to alleviate guilt or satisfy a need to know that everything was done to save their parent, sibling, or close relative who had been neglected or not given adequate care by the very concerned relatives.

The solution requires the intense education of the public, its elected government representatives and health professionals addressing over utilization of patients and the limitations of the present day modern medicine in the treatment of the aging/failing body.

Over-utilization by Healthcare Professionals

Physicians and other healthcare providers play an important role in the healthcare expenditures of our nation. Physicians have the tendency to use the newest drugs, therapeutic procedures, and technologies available in the care of their patients. Perhaps the reason is their sincere desire to make the right diagnosis as quickly as possible, so effective treatment may begin immediately. This desire is ingrained in their medical education, post-graduate training, and continuing medical education, and becomes second nature to physicians.

In addition there is always that lingering thought of a missed diagnosis that leads to a possible malpractice suit if the best and latest available diagnostic procedures and therapy are not employed; especially among patients knowledgeable of their condition and at the same time requesting for the most advance medical management. To protect themselves from liability suits and appease their patients, physicians and other health providers often comply resulting in excessive and unnecessary medical procedures that drain the healthcare system of its available funds.

These problems can be easily corrected by educating, retraining, and encouraging physicians and other health providers to practice **Clinical Evidence Based Medicine and recognized Institute for Clinical Systems Improvement (ICSI) Guidelines**—a comprehensive treatment guideline initiated and sponsored by the major healthcare institutions/players in the state of Minnesota—or any other **Physician's Clinical Practice Guidelines** for every specialty and integrative medicine that may be nationally developed, standardized and adopted later on.

An axiom most U.S. physicians observe and subscribe to is, *"Medical decisions must be based not on cost or any other consideration but what is best for the patient."* In most cases these will mean the use of aggressive, intensive, invasive and very expensive medical care decisions. This idealism is great in a perfect world. We live, however, in an imperfect world with many unforeseen variables. The indiscriminate practice of this motto unfortunately sometimes leads to many unpredicted immediate and/or long term consequences that are equally important to the patient, her/his family, and the community where the patient lives.

Physicians who adhere to this medical "golden rule" are well advised to practice it with great caution, making appropriate and carefully thought-out medical decisions; remembering that there are many well-designed studies indicating **more care does not** *necessarily equate to better care. In most cases, more care means higher mortality and morbidity rate.*

Another area of abuse by physicians and other healthcare providers is the way professional fees are structured. There is great variation in fees from doctor to doctor of the same specialty, specialty to specialty and one location to another. The prevailing methods by which healthcare providers (be they doctors, ancillary healthcare providers, clinics, hospitals, etcetera) charge their services or products have no rational or logical basis. The result is the frequently observable abuses. How can one explain a dose of aspirin or Tylenol in the hospital costing ten dollars? Why are there wide differences in pricing of the same drug between the lowest and the highest price which at times is up to 600 percent from one pharmacy to the other?

A good solution to this problem is to standardize fees and prices for healthcare products. The basis of a fee or price must be defined and determined. Once the parameter for a fee or price is established each particular healthcare service or product must be assigned a Relative Value Unit (RVU) which then becomes the basis of pricing and reimbursement.

Such an RVU system is already in existence and used by many healthcare practitioners. The present scope of the RVU is, however, limited and must be reformulated and expanded to include all foreseeable types of healthcare services and products in the future. Only then can it be implemented and becomes universally acceptable.

Profit-motivated Companies

Profit motivation—either by **for-profit** or **not-for-profit** healthcare organizations—is another significant driving force for the escalating medical care costs. Despite recent economic downturns, healthcare organizations are reporting record profits. CEOs in the healthcare industry are compensated up to $100,000,000 a year, and their immediate high level colleagues and administrative staffs are not far behind. Although most healthcare management teams claim a healthcare administrative overhead of 10 percent to 15 percent, recent studies by Harvard Medical School and Public Citizens account 31 percent or $399,400,000,000 annually of the total U.S. healthcare spending to bureaucracy. In Canada it is 16.7 percent. Added to legitimate administrative expense are expenses that have nothing to do with direct patient care. These non-direct healthcare expenses are often labeled as capital expenditures. This is where the profits generated are directed and hidden. Such an accounting gimmick allows healthcare companies to expand their real estate investments in offices, healthcare facilities, the latest in medical equipment, acquisition and merging with other companies, and other health-related businesses. There are also the pharmaceutical companies selling their products in the United States at three to four times more than in other developed countries—to recapture expenses in R and D as their main argument.

Yet they spent billions of dollars in advertising, inciting public consumers to demand the newest and most expensive drugs available from their healthcare providers. It was only in the last decade that prescription medications are advertised. In the post-WWII era to the early 1990s prescription drugs were not advertised. Again as soon as drugs were openly advertised healthcare expenditures accelerated. Prescription drugs had been one of the areas of great concerns because of rising medication cost, particularly to the seniors. No wonder health insurance premium and medical care are continuously going up. This is what happens when healthcare is treated just like an ordinary business and not as a profession where patient care is the primary objective and profits a distant secondary goal. Perhaps taxation is the answer for profits channeled to investments and other business activities that are not related to direct patient care.

A very effective solution to this problem is the adherence and enforcement of the principles of fair competition, and the allowance of the free market dynamics to work with the least interference from the government. This implies that companies unable to compete must not be helped or bailed out financially by the federal or state government—no matter how important or powerful they may be; no matter how bad the situation is. Even a company that is "too big or to influential to fail" should be allowed to fail. This means also a reexamination of our federal/state laws related to company merger/takeover, and the stricter and tighter enforcement of the U.S. monopoly laws. The two-step bidding process of this proposed healthcare initiative is perhaps the best means in controlling the rise in healthcare cost.

Malpractice and Litigation

Malpractice is a constant concern of healthcare practitioners because of the ever increasing multimillion dollar awards juries across the United States give to the plaintiffs. In this era of litigations that can potentially ruin careers, finances, and lives, who can blame physicians and other healthcare professionals for ordering the newest and most expensive tools at their disposal, indicated or not, to cover themselves in the event of future malpractice litigation? Failure to order expensive state-of-the-art diagnostic and therapeutic procedures is misconstrued as negligence and/or incompetence in a malpractice suit. The end result is too many unnecessary tests and treatments are performed that escalate healthcare cost and malpractice insurance premiums. Due to the unreasonably high and unaffordable premiums many doctors forego their malpractice insurance coverage, retire, or quit doing high-risk procedures and treatments in obstetrics, orthopedics, neurosurgery, cardiovascular, radiation and cancer therapy, etcetera.

The real questions are: "Why are medical injuries (medical accidents) sustained in the course of medical therapy treated differently from occupational or motor vehicular accident injuries? Is it because health professionals have high insurance liability coverage? It is of interest to note that most malpractice suits where multimillion dollars are awarded occur in exceedingly specialized medical fields performed by highly trained and well-qualified specialty doctors or healthcare providers. Maybe society's expectation of modern medicine is unjustly considered so perfect that no unexpected outcome is tolerated.

The solution is **tort reform** limiting medical malpractice liability to a reasonable and fair compensation for the victim. A maximum award of $3,000,000, for example, plus predetermined legal fees for injuries resulting in death or permanent total disability to the victim is fair and reasonable. Other forms or degrees of disabling injuries resulting from unintentional medical complications, injures, unexpected treatment outcomes or negligence may be rated according to a system similar in principles with the ***Workman's Compensation Disability Rating Protocol (WCDRP)***. This can be referred to as the ***Healthcare Malpractice Compensation Protocol (HMCP)***. It is recommended that such a protocol is to be federally mandated. It is a fair and reasonable means of compensation if a universal healthcare plan is passed since the expenses for future medical needs of the injured victim is no longer an issue or concern.

Ratings in the HMCP for partial permanent disability may be expressed as percentage of loss of function of the injured person or body part similar to the WCDRP percent rating of body parts. Compensation for partial permanent disability ratings must be proportional to the maximum allowable compensation of say $3,000,000. For pain and suffering a maximum award of $250,000 is reasonable.

Tort reform and malpractice awards modifications should be instituted. Arbitration to settle disputes should be the first step before any law suits and court proceeding.

Defensive Medicine

In the course of a normal patient examination and diagnosis, doctors most of the time do not order additional test to make a diagnosis, especially when the patient is well known to the doctor. In more complicated cases doctors order diagnostic procedures they believe are appropriate. When the patient and/or close relatives ask for more diagnostic tests, doctors take the time to explain why such tests are unnecessary. Most of the time, the patient will accept the doctor's explanations. Many, however, will insist in having more expensive up-to-date diagnostic tests done and/or the latest treatments available. To refuse the demanded test means time consuming explanation, irate patient, and a likely lawsuit in the event a remotely possible diagnosis happens to be the case. The same is true for treatments. Many patients demand only the best, latest, and most expensive cures available.

It is not unusual therefore, for doctors and other health practitioners to order unnecessary diagnostic procedures, and treatment programs in anticipation of a possible litigation years later by the demanding patients or relatives. This is known as

Defensive Medicine. There are many estimates as to the cost of defensive medicine.

In 2006, Pricewaterhouse Cooper LLP reported about 10 percent of medical cost is related to malpractice lawsuits. Quoting Jackson Healthcare CEO, Richard L. Jackson, in a speech delivered at the National Press Club in Washington (2010) on defensive medicine: *"$1 in every $4 spent on healthcare each year is spent on unnecessary tests and treatments ordered by physicians solely to protect themselves against lawsuits."*

A revised statement by Mr. Jackson later reported by Rene Letourneau of Healthcare Finance News in 2010, estimated defensive medicine at a price tag of 26 percent to 34 percent, or $650 billion to $850 billion of the $2.5 trillion national health expenditures.

The reason for defensive medicine is medical malpractice lawsuits. Majority of medical malpractice lawsuits will settle out-of-court; with an average out-of-court settlement of about $500,000. Juries find hospitals more frequently liable for malpractice than doctors. Statistics show 50 percent of hospitals sued and 33 percent of lawsuits against doctors are successful. Additionally, the average hospital medical malpractice lawsuit pays $6,000,000 to $10,000,000, while a malpractice suit against a doctor is about $2,650,000 to $4,500,000. On Feb. 2, 2003, a Norwalk couple was awarded $58,600,000, a record for a single incident of medical malpractice in Connecticut. The case involved an obstetrician accused of waiting too long to perform a cesarean section, and the child was permanently brain-damaged with severe cerebral palsy.

Studies show that 25 percent of all practicing physicians are sued yearly; and 50 percent to 65 percent of MDs in practice are sued during the entire period of their professional careers.

Fear for medical malpractice is the main driver for defensive medicine. A medical malpractice lawsuit against a physician can mean bad reputation, or completely ruins a doctor financially. **Tort reform** *is the only solution to stop this madness.*

New Technological Advances in Healthcare

The explosion of new technologies in the treatment and diagnosis of human diseases has been tremendous in the last fifty years. These advances in medical science are the product of basic scientific researches encouraged and funded by the federal government with American tax monies. The resulting scientific findings are then handed to commercial institutions that do further research, and develop products that become new drugs, therapy, diagnostic instruments, biotechnology, bioengineering, etcetera.

The new products—later developed from these basic researches by commercial firms—are then sold to Americans at exorbitant cost. The reasoning, which seems logical and equitable, is the recovery of research and development (R&D) expenses. The fallacy of the argument is the charging of Americans double or triple the price of what is charged in other countries, e.g., Canada, European countries, Japan, Korea, etcetera, when, in fact, it was the American taxpayers who financed the initial research of such products. The burden of shouldering R&D expenses should be equally distributed among citizens of other rich and economically developed nations so Americans may experience some reductions in the high cost of their medical care.

A logical proposal is to negotiate for a fair and equitable compensation from private companies to pay royalties to the U.S. government for any product developed later from such basic researches, and to sell such products to Americans at a reasonable price similar to other countries.

A less favored solution is for the federal government to regulate the cost of new technologies in the healthcare industry.

Illegal and Non-permanent Resident Aliens

Illegal and non-permanent resident aliens are siphoning billions of our healthcare dollars. Many border states, like California, Arizona, New Mexico, and Texas, report spending billions of dollars taking care of illegal aliens in their community, city, and county hospitals and clinics with nobody taking the responsibility of paying for their medical bills. Uncompensated emergency room and hospital services given to illegal immigrants had led to severe financial strain to some hospitals, that in California alone between1996 and 2004, eighty hospitals closed as a result. Ninety-five percent of ER physicians give free medical care and in LA County alone in 2000 to 2002 $167,000,000 was lost from unpaid medical bills made available to undocumented immigrants. From the year 2000 to 2005, the cost of treatment for illegal immigrants (mostly given in emergency rooms) increased immensely and continues to soar. In California—home of 40 percent of the estimated 12,000,000 illegal immigrants—taxpayers have paid an average of $957,000,000 annually since 2001.

In other states, the healthcare expenditures for undocumented immigrants revealed for the state of Georgia from $58,000,000 in the year 2002 to $112,000,000 in 2005; Kentucky, from $2,200,000 in 2003 to $9,000,000 in 2005; North Carolina doubled from $25,800,000 in 2000 to $52,800,000 in 2005, and Minnesota's spending jumped from $12,500,000 in 2003 to $17,400,000 in 2005—a 40 percent increased. Other states with large illegal immigrants, i.e., Arizona and Texas, have no data as to how much they spend for the medical care of their illegal immigrants. But in Houston (Harris County Hospital District) the spending in 2005 was $97,300,000, which represented 14 percent of the annual operating expenditures. This results in depriving these states and cities across the United States of much needed billions of healthcare dollars for their citizens. Sometime in the middle of 2004, national TV news stations reported that in the east coast, there were many cases of Middle Eastern women coming especially for medical treatment and to deliver their babies paid by the state medical assistance programs. They then go back to their countries and leave their babies with relatives to be supported by the state welfare and Medicaid programs.

These are not isolated cases. It has become rampant and is happening at an alarming rate all over the United States. Citizens of other countries are taking advantage of our generosity and openness. Our porous borders and unenforced immigration laws encourage illegal migration to our country, compromising our own security as a nation. Many Americans, who may be rich, influential and political activist, are assisting this, as a result of it being a source for cheap labor or due to a real concern for the plight of poor people everywhere.

The solutions are: 1) strictly enforce our immigration laws, 2) reform our immigration laws to allow foreign scientists, professionals, and other workers easy and timely access to meet and invigorate our economic needs, 3) tighten up and regularly patrol our borders using our National Guards with each state National Guard rotating to cover the borders in need of securing, 4) issue a National Healthcare Identification card as proposed, and 5) bill foreign governments for the medical care of their citizens who are illegally in the U.S. or who are legal non-permanent resident aliens in the U.S.

Solution number five may seem outrageous. In most countries with national healthcare systems, however, foreigners are required to provide their own healthcare coverage and are responsible for their healthcare expenses. Very few countries will absorb foreigner's healthcare expenses while in the country.

Mechanisms on how to collect unpaid medical bills by illegal and non-permanent resident aliens to respective foreign governments (embassy) must be included in the national healthcare plan. Non-payment of a foreign country can lead to a freeze or lien of the foreign government's assets in the U.S.

Capping or Limitation of Medical Benefits

To insure the financial solvency of a national or universal healthcare plan for decades to come, hard choices must be made and should be included in the planning of the system. Obviously, no matter how well financed a healthcare system that proposes to cover every citizen of a country, breakdown of the system (sooner or later) is to be expected from financial burden and unexpected expenses. It is in this regard that capping or limitation of healthcare benefits be defined and incorporated into any national healthcare plan to be developed.

Capping or limitation of medical benefits does not mean the withdrawal of medical care from a patient. It means the use of proven and equally effective but conservative healthcare management. It is the avoidance of unnecessary, aggressive, expensive, state-of-the-art medical and surgical treatments, which by many indications cause higher mortality and morbidity rates. These conservative treatment approaches can be easily implemented using ***Healthcare Teams (HCT).***

A Healthcare Team is a group consisting of healthcare providers with various expertise; from other disciplines of the medical fields working together to attain the specific purpose of the HCT. Depending on the goal and nature of the HCT, the members of the team may include a physician, nurse practitioner/specialist, dietitian, social worker, pharmacist, exercise physiologist, physical therapist, psychologist, medical ethicist, alternative healthcare practitioner, community organizer/worker, etcetera. As an example: a Diabetic

HCT would have a physician, diabetic nurse specialist, dietitian, podiatrist, pharmacist, and perhaps an exercise physiologist. In every specialty of medicine, one can organize a specific HCT that will be responsive to the needs of the patients with the diagnosis. The HCT can be in-hospital, outpatient, or community-based.

As an example, a HealthPartner clinic started a Cardiac HCT. The members were an MD internist, cardiac nurse specialist, dietitian, social worker, pharmacist, and exercise physiologist. Heart patients were enrolled and were allowed to attend the Cardiac HCT. The Cardiac HCT was conducted, once a week as a regular clinic after hours, for one to two hours. Like any clinic, the HCT clinic conducted adjustments or changes of medications and other treatments. It was later moved on Saturdays from 8 am to 12 noon to accommodate more patients.

The Cardiac HCT clinic started as initial question and answer session. The Q&A was directed to any member of the HCT. The patients were then allowed to visit with as many HCT members of their choice to discuss their issues and concerns. The program was successful. The patients liked it. The HCT can be designed around the convenience and needs of the patients. Patients in the Cardiac HCT program were also followed by their attending/cardiologist physicians on scheduled follow-ups, as seen appropriate by both the doctor and patient.

There is no easy solution to this dilemma. Studies show that during the last six months of a person's life, the most intensive care is given. It is during this time aggressive and expensive medical decisions are made. The "book" is literally thrown to the patient. *Every possible diagnostic and treatment avenue is explored and pursued. It can not be over emphasized that **more care at times is harmful** especially in the elderly, and 5 percent of Medicare patients who would live three to four months spend 60 percent of Medicare financial resources.*

Fraudulent Claims

This is another area where billions of healthcare dollars are lost, and should be looked into critically. An estimated $100 billion to $125 billion of healthcare dollars were lost in 2010 by fraud. Fraudulent claims take the form of billing-without-service to accounting schemes practiced by some healthcare providers from the individual practitioners to the largest healthcare institutions. Unless these abuses are taken seriously and stopped, this area of financial lost will continue to grow as a significant factor in healthcare expenditure.

The solutions are to treat fraudulent claims as a serious federal and/or state offense, punishable by fine and mandatory imprisonment, and periodic surprise federal/state governmental audits of large healthcare institutions—similar to the audits at Allina, Blue Cross Blue Shield of Minnesota and HealthPartners done by the Office of the Attorney General of Minnesota, Mike Hatch in the early 2000.

Lobbying and Politics

To lobby is to press for political action in favor to the lobbyist's advocacy, which most of the time, protects the profit, or increases the income of the lobbying group.

There are many factors driving the fast growth of healthcare cost in the U.S. Many reliable sources agree lobbying is a major factor. There are, however, no reported estimates found that confirm directly the effect of lobbying on healthcare costs. On the other hand, it is safe to assume that lobbying increases the costs of healthcare.

In January 2013, Fox New reported the nature and effect of lobbying on business outcomes. In 1973, only 14 percent of corporate boards have well connected ex-government officials in their boards. In 2012, 50 percent of big U.S. companies have well known ex-politicians as members of their board of directors. The reasons are obvious. These past Washington politicians are paid handsomely and given company stock options, and other benefits. They have many friends, contacts, and personally know a lot of their former colleagues still serving Congress and the Executive branches of government. They are experts in maneuvering the inner workings of Washington D.C.; making them truly effective and influential lobbyists.

According to two University of Kansas professors, lobbying is a lucrative business, better than introducing a new product or service. For every $1 invested in lobbying comes back $220, or 22,000 percent rate of return. No wonder G.E. spends $107,643 per day, 24/7 for lobbying. The ex-government official board member of G.E. usually targets members of Congress who acted favorably for G.E. in the past according to the report.

In 2011, large health insurance and pharmaceutical companies, hospitals and clinics, medical professional groups, unions, and so forth spent $1.2 billion lobbying the U.S. Congress. A good example is the $40 billion per year Medicare D program for seniors. This program prohibits price negotiation with the pharmaceutical industries. The ObamaCare according to projections will increase healthcare cost by12 percent; surely is another example. My guess is that, lobbying increases healthcare cost by about the same 12 percent as estimated in the ObamaCare.

There is no question that lobbying and politics play important roles in the healthcare system of our nation. Multi-million/billion dollar corporations—especially in the health insurance and pharmaceutical industries—give large amount of political contributions to effectively influence elected officials. Politicians on the other hand welcome and willingly accept monies for their campaign funds and other forms of gifts, and later claim the decisions they make—usually in favor of the lobbyist—were not an influence or a factor in their decision making. Do you think so? The recent political upheaval in Washington, D.C. brought by the Jack Abramoff lobbying scandal illustrates well what one influential lobbyist can do swaying members of Congress to act in favor of a lobbyist's political agenda.

The end result of politicking and lobbying is usually the maintenance of the status quo or worsening of the problems and situations in any industry where lobbyists are actively engaged, and attain their goals. Some people would argue that lobbying is a form of expression of the "freedom of speech" guaranteed by the U.S. Constitution, and therefore can not be effectively stopped. If so, then the U.S. Congress is obligated to rewrite the present lobbying rules and even out the playing field between the rich and famous and influential large corporations on one hand, and the common people on the other hand.

A reasonable solution is to change and introduce new lobbying laws that will effectively thwart the influence peddling of the rich and famous, and big, powerful corporations. Another is to establish a permanent federal commission that constantly monitors the activities of the lobbyists and reports to the general public, especially to the elected official's constituents on how they voted on important legislations related to a lobbyist's agenda. Failure of the federal government to establish a permanent commission with the only purpose of monitoring lobbyists and elected officials voting records is no excuse not to have such an advisory group. A not-for-profit association can be organized, especially by consumer advocacy groups funded by voluntary public contributions. It may be called *Congressional and Lobbyist Watchdogs*. The primary objective of the group is to promote "clean" transparent government, through the constant monitoring of activities of the powerful and influential civilians and government officials, with the hope of preventing graft and corruption.

Isolated Public from the Actual Cost of Healthcare

Since the advent of Medicare, Medicaid, employee's medical benefits, and health insurances, the public had been increasingly isolated from the actual cost of their medical care. Patients in this country personally do not have any idea how expensive medical care has become. Most of the time what they see is the amount of their co-pay and/or deductible representing a very small fraction of the actual expenditure. This is usually shrugged off and an attitude of indifference prevails. In countries like France, Germany, and Switzerland, the cost of medical care is easily visible in payment receipts issued by providers. A very high percentage of the population is aware of the actual cost of her/his medical care, and consciously becomes prudent in accessing healthcare benefits, particularly so when one is asked to pay medical bills at point-of-service.

The solutions are: 1) to develop a system that can automatically determine the actual cost of the office medical care visit; 2) to require healthcare providers to issue patients receipts showing the actual medical expenses; 3) to have the patients pay the amount they are responsible for at-point-of-service.

Health Attitude of Americans

Young Americans are likely not to pay too much attention to their health. They tend to ignore simple common sense preventive measures to maintain realistically good health. To most young people regular exercise, healthy eating habits, rest and relaxation, and a healthy lifestyle are not in their vocabulary. They feel invulnerable. They overeat, smoke, abuse alcohol, use drugs, practice indiscriminate sex, work too much, and socialize until the early morning hours, sleep deprived, stressed, etcetera, that as they enter their fourth or fifth decade of life, they have already acquired the pathological effects of lifetime abuses to their health. Unfortunately, many Americans continue to practice unhealthy lifestyles in spite of serious medical conditions they may have. Only a few exceptional individuals are able to change to a healthier lifestyle for good. Lifestyle is hard to change. Habits are difficult to break.

Early preventive education in primary through secondary school must be a continuous major part of the school curriculum just like reading, writing, arithmetic, natural science, social studies, history, and so on. This series of health education may be labeled **Health Science Studies.** *Its main purpose is to inculcate to young American minds the importance of healthy lifestyle.*

Cumbersome Paperwork

Medical practice in the present age requires proper documentations of all business transactions and patient encounters. From the time a person enrolls for insurance coverage paper records are created, which in turn are regularly updated. Healthcare providers, be they physicians, dentist, or practitioners of integrative medicine, clinics, hospitals, etcetera, submit bills generating more paperwork. The problems of paperwork are further magnified by the requirement and insistence of the different governmental agencies and private healthcare institutions to use their own individual forms, causing an insignificant part of the business of healthcare to be complicated, time consuming, wasteful, inefficient, and expensive. On the patient encounter side of the spectrum, every patient visit to the doctor requires again the updating of personal information and signing of documents for insurance, billing, and permission to be treated. The doctor's encounter with the patient has to be precisely documented as future reference for the patient's medical care follow-up and to be properly compensated for the services performed. More paperwork is produced requiring labor-intensive staff handling and processing.

To improve and gain easy and simple access to medical care and patient's records, the cumbersome use of paper billing and medical record keeping must be eliminated, and converted into electronic form. ***Electronic Medical Record (EMR)*** keeping or ***e-Health*** as some Europeans countries call it, is fast becoming a standard of operation in the healthcare industry, at least in the state of Minnesota. With a "click of the mouse," a healthcare provider can easily document the patient meeting with the provider in the format of ***Symptom, Objective findings, Assessment and Plan of care (SOAP)***, order laboratory/diagnostic tests including sophisticated procedures, refer patients to medical consultants and prescribe or refill medications to a variety of pharmacies chosen by the patient. The best part of EMR is the availability of the patient's entire medical record, literally at the fingertips of the provider, which can be instantly reviewed without difficulty, improving medical care.

With increasingly widespread use of the EMR/e-Health format, not only in the United States but also around the world, coupled with the globalization of commerce, and growing mobility of populations, it is vital for medical communities around the globe to contribute and cooperate in the development of an encompassing standard system of medical and other healthcare-related terminologies easily used in administrating and conducting healthcare systems here and abroad. Such an international medical standard of healthcare nomenclature must be universally understandable, recognizable and easily applicable.

The momentum to create an integrated EMR/e-Health system here in North America and Europe by government-promoted programs is important to all those who will actively participate and use such a system. Here in the U.S., the Secretary of the United States Department of Health and Human Services last May 2004 announced the appointment of the ***National Health Information Technology*** Coordinator to direct the fulfillment of President Bush goal of an EMR for most Americans by the year 2014. Secretary Thompson declared: "Health information technology promises huge benefits, and we need to move quickly across many fronts to capture these benefits." He stated this can be accomplished by establishing secure local, regional and/or national computer networks that "would allow a doctor or healthcare provider to access an always up-to-date electronic health record of a patient who has agreed to be part of the system, regardless of when and where a patient receives care."

In the U.S. and U.K. attempts to address the deficiencies of the existing medical system of terminologies has been ongoing for the last several decades. Efforts from these countries led to the development of the *Systematized Nomenclature of Medicine (SNOMED)* in the U.S. and the Read Codes—later known as *Clinical Terms (CT)* from U.K. The two later were integrated and merged to form the *SNOMED CT*—an extensive amalgamation of more than 300,000 distinctive healthcare concepts and more than 1,000,000 descriptive terms. In both the U.S. and U.K. this combined system is still being worked out and refined. Hopefully an internationally acceptable and implementable standard medical nomenclature is soon created. Similarly, confusion in the names of prescription drugs has to be eliminated. A drug should be marketed with a single brand name worldwide or by its generic name so healthcare providers anywhere in the world will immediately recognize and know what the drug is for.

Imagine being admitted in any hospital in the United States or in any country where EMR is in use. By a single swipe of a healthcare ID card, providers can instantly access your follow-up cares or medical records and swiftly track down your billing information; making the system easily, efficiently, and effectively administered.

Addressing all the above problems identified will definitely and significantly reduce healthcare costs in the United States of America.

IV

Patient Protection & Affordable Care Act (Obamacare)

The ***Patient Protection and Affordable Care Act (PPACA/HR3200)*** signed into law on March 23, 2010 is better known as the ***ObamaCare***. It is not a universal healthcare plan. The original intent was to address the 47,000,000 Americans without healthcare insurance. Since then, however, HR3200 has slowly changed. As more regulations are added ObamaCare is becoming a heavily regulated government healthcare system, similar to its counterparts in other countries. With certain provisions in the law, I would not be surprising if in the next few years, HR3200 will slowly evolve and take over the entire healthcare system of this country; transforming it into a universal healthcare plan that is single-payer government controlled like those of Britain, Canada, and Sweden.

Since the election of President Barack H. Obama in 2008, healthcare became one of the hottest issues of concern both by the general public and the government. Unlike President Bush, one of President Obama's major goals was to sign a national healthcare bill as his legacy. Perhaps he realized that left alone healthcare cost will continue to skyrocket adversely affecting the people, economy and the country as a whole.

With the economic crisis, the country has, President Obama was adamantly pushing through his national healthcare agenda. Unfortunately, he did not give the U.S. Congress what specific areas in the present healthcare he wanted reformed. Instead, he gave Congress vague principles and allowed members of his democratic party who were the majority at the time, to draft and legislate a healthcare bill. After one year in office, President Obama did not get the healthcare bill passed.

America's Affordable Health Choices Act of 2009 (HR3200) was introduced in the 111th Congress of the U.S. House of Representatives on July 14, 2009. The bill was superseded by a similar bill, the Affordable Healthcare for America Act (HR 3962), which was passed by the House in November 2009. On September 17, 2009 a similar bill called the "Affordable Health Choices Act" (HR1679) was introduced in the Senate. The Senate instead passed another version called the "Patient Protection and Affordable Care Act."

Congress in both the House of Representatives and the Senate could not agree. The House bill had about 1,990 pages and the Senate bill was 2074 pages. There were many back and close door negotiations mostly among the Democrats. Republicans where not included. Squabbling by members of the lower house and senate, accusations and counteraccusations were intense and frequent. In the end, no one knew exactly what were in both the House and Senate bills. Attempts to reconcile the healthcare bills were unsuccessful and the healthcare bill failed to passed and tabled. The two bills were reconciled anyway and pushed for a vote. Quoting the Speaker of the House Nancy Pelosi, ***"But we can pass the bills ... so that you can find out what is in it ... away from the fog of the controversy."***

No one in both Houses of Congress read the reconciled bills. The Representatives and Senators did not have the time to study them. Copies of the reconciled bills were distributed to members of Congress less than 36 hours before voting. The Democrats were the majority in both Houses of Congress, and the Affordable Healthcare Act was finally passed. The final healthcare reform bill was estimated at 1,074 pages called the Patient Protection and Affordable Care Act (PPACA/HR3200). In short, it is sometimes called the Affordable Care Act. More popularly it is known as ObamaCare, which the President happily and proudly embraced recently.

Major concerns and criticisms on both sides of the aisle in the U.S. Congress are the projected expense in the trillions of dollars; the taking over of the government of the healthcare system which represented 1/6 of the nation's economy ($2.52 trillion); the lack of distinct ways of financing the new health bill; concerns about the government becoming the single–payer, inefficiency, long waiting and rationing of healthcare.

On March 23, 2010 President Barack Obama signed into law the Affordable Care Act or ObamaCare. Because of controversial key elements of the law, several states questioned the constitutionality of the reform healthcare act. On June 28, 2012, the Supreme Court upheld the law as constitutional. The individual mandate to have a health insurance was interpreted by Chief Justice John Roberts as a tax and not otherwise. In so doing, he sided with the four liberals in the Supreme Court. He became controversial overnight. Nonetheless, the Affordable Care Act (Bill Number HR3200) became the law of the land. Conservative Republicans were furious and vowed to repeal the law if they win the President, House of Representatives and Senate in the 2012 election.

Despite these controversial issues, ObamaCare became the healthcare law of the United States of America. Many Americans supported it and communicated it to their elected officials in Washington, D.C. There are probably just as many who opposed HR3200. As we know more about ObamaCare more seems to oppose it, because of certain provisions of the law. Here are some aspects of ObamaCare reviewers find very positive:

- Pre-existing conditions of children will no longer be a reason for health insurance companies to deny health coverage to a child. By 2014, adults will qualify for the same protection as a child with pre-existing illnesses.

- Health insurance companies can not drop your healthcare coverage if you become sick.

- It allows parents to include in their healthcare insurance coverage children up to age 26 years old.

- It can not modify and/or limit the coverage you are entitled to over your lifetime.

- In 2010, you may be entitled to a rebate for overpayment from your insurance company.

- PPHCA obligates health insurance companies spend at least 75 percent of insurance premium paid on medical services, and not on any other administrative expenses, e.g. advertising or lobbying.

- Healthcare companies are required to submit rate hike justifications to the states.

- You may not pay a co-pay for preventive/wellness care and pregnancy exam.

- Health insurances in state-run exchanges will enable you to compare and shop for the health plans that best suit your needs. These exchanges will help you find other government health benefits, you are entitled or if you qualify for tax credits.

- In 2014, if your income as a single is under $14,000 and $29,000 for a family of four, you may qualify for Medicaid enrollment.

- If your income is under $43,000 for single and $88,000 for a family of four, you can apply for tax credit on your healthcare expenses, and reduced co-pays and deductibles.

- In 2010, if you have Medicare Part D and in the "donut hole" you may receive $250 cash rebate. By 2011, you are eligible to get discount on all brand-name and generic prescription drugs.

- Employers are allowed to discount premiums for employees who participate in wellness programs and achieve certain health goals.

- For small businesses, ObamaCare calls for the owners to provide health insurance for 50 employees or more. However, if there are less than 100 employees, the owner may participate in state-run exchanges of their choice, in 2014—a cheaper alternative than what is presently available. Any employer who refuses to provide health insurance to their employees is penalized a fine of $2,000 per employee, with the exception of the first 20 employees.

- Small businesses with 25 employees or less, who provide insurance for their 25 employees or less, qualify for a tax credit up to 50 percent in 2014.

On the other hand, here are strong arguments as to why the ObamaCare should be repealed. No wonder our nation is a divided nation when it comes to the Affordable Healthcare Act.

- ObamaCare does not address the various problems that continuously accelerate the cost of healthcare in America.

- Douglas Holtz-Eakin's (former Congressional Budget Office head) analysis of the HR3200 law, found that, if one accounts for all the factors CBO could not, then ObamaCare would very well increase the deficit by $190 billion.

- The analysis of the chief actuarian of Medicare, Richard Foster's showed that ObamaCare would not reduce healthcare costs; rather it would increase by about $311 billion through 2019.

- ObamaCare contains the largest hidden tax hikes imposed on the American economy; estimated at about $800 billion over a decade by the Congressional Budget Office. By 2035, these initially unrecognized tax increases are expected to add another 1.2 percent of GDP.

- It will exacerbate the nation's unwarranted entitlement spending, particularly on Medicare and Medicaid, and possibly push the federal budget to the breaking point.

- The ObamaCare was supposed to cut about $200 billion from Medicare Advantage—the Medicare program that pays private insurers to provide Medicare benefits—to help fund the HR3200 law. Eleven million seniors with Medicare Advantage plans, if the cuts were enforced, could lose their plan. The Obama administration instead suspended the cuts for 2011 and adjusted by 1.6 percent to increase in 2012 election year.

- A provision of ObamaCare compels all state to expand their Medicaid program to the federal poverty level. This is another added cost to states, estimated at $118 billion by 2023. Medicaid already burdens states over budgets.

- The seventy-five percent medical-loss-ratio (MLR) regulations demand that 75 percent of premiums have to be spent on patient care. The remaining 25 percent is for administrative and other expenses. MLR is an excellent reason why 36 states are opting-out of the ObamaCare.

- ObamaCare mandates every transaction worth $600 done by healthcare businesses to submit tax form 1099 to the IRS—a huge requirement that means costly new paperwork.

- Greg Scandlen discovered ObamaCare Basic Health Plan "states may implement to provide coverage for people between 133 percent and 200 percent of poverty to non-citizen legal immigrants who are not eligible for Medicaid."

- Section 2711 waivers permit businesses, labor unions and other groups to opt out of ObamaCare; and so far the DHHS has granted 1,372 waivers.

- Health insurers are abandoning the child-only market because the pre-existing conditions no longer are a factor in denying coverage. The other reason is the inclusion of children up to age 26 years old in their patient's health plans.

- The controversial life after death panels aroused from a provision of ObamaCare allowing Medicare to pay end-of-life counseling; a decision to be made by Washington, D.C. faceless bureaucrats which Gov. Sarah Palin called "death panels."

- A major source of innovation in healthcare is "physician-owned specialty hospitals." However, ObamaCare prohibits the establishments of new physician-owned specialty hospitals and makes it almost impossible for existing ones to expand.

- It handed over to DHHS to control just about every aspect of the nation's health system.

Although the final reconciled bill was 1,074 pages, several months later it grew to 6,000 and the latest count in 2012 is 13,000 pages as more regulations and amendments are added. There are still many sections and provisions not written yet. So far, the drafting of additional regulations continues. ObamaCare, as it is, has already touched every aspect of medical practice. The government heavily regulates it in particular by the Department of Health and Human Services (DHHS). An overview of HR 3200 is included to give the reader inkling on how intricate and extensive the law is. The general outline gives you an understanding how Congress dealt with the complexity of the Patient Protection and Affordable Care Act. The sections selected should give the reader a fair insight on what this healthcare bill covers. The sections are the general items in the bill. Please keep in mind that in each section are numerous subsections, subtitles and subtopics. They are the main topics of discussion and are divided into more detailed provisions under subsections, subtitles and subtopics. It is, therefore, not unusual for a single section to cover pages upon pages of provisions in the law. For more details of the Affordable Care Act, please visit the following websites where most the laws were taken:

http://www.govtrack.us/congress/bills/111/hr3200/text
http://thomas.loc.gov/cgi-bin/query/z?c111:H.R.3200

H.R. 3200 (111th): America's Affordable Health Choices Act of 2009

111th Congress, 2009–2010. Text as of Oct 14, 2009

H.R. 3200 Congressional Outline

I.   SECTION 1: SHORT TITLE; TABLE OF DIVISIONS, TITLES, AND SUBTITLES.

    A.   DIVISION A—AFFORDABLE HEALTHCARE CHOICES

        TITLE I—PROTECTIONS AND STANDARDS FOR QUALIFIED HEALTH BENEFITS PLANS

        TITLE II—HEALTH INSURANCE EXCHANGE AND RELATED PROVISIONS

TITLE III—SHARED RESPONSIBILITY

TITLE IV—AMENDMENTS TO INTERNAL
REVENUE CODE OF 1986

B. DIVISION B—MEDICARE AND MEDICAID
IMPROVEMENTS

TITLE I—IMPROVING HEALTHCARE VALUE

TITLE II—MEDICARE BENEFICIARY
IMPROVEMENTS

TITLE III—PROMOTING PRIMARY CARE,
MENTAL HEALTH SERVICES, AND
COORDINATED CARE

TITLE IV—QUALITY

TITLE V—MEDICARE GRADUATE MEDICAL
EDUCATION

TITLE VI—PROGRAM INTEGRITY

TITLE VII—MEDICAID AND CHIP

TITLE VIII—REVENUE-RELATED PROVISIONS

TITLE IX—MISCELLANEOUS PROVISIONS

C. DIVISION C—PUBLIC HEALTH AND WORKFORCE
DEVELOPMENT

TITLE I—COMMUNITY HEALTH CENTERS

TITLE II—WORKFORCETITLE

TITLE III—PREVENTION AND WELLNESS

TITLE IV—QUALITY AND SURVEILLANCE

TITLE V—OTHER PROVISIONS

H.R. 3200 Sections (copied with slight modification of the bill's language).

Sec. 100: ***Purpose: Table of Contents of Divisions; General Definition.***

The purpose of this division is to provide affordable, quality health care for all Americans and reduce the growth in healthcare spending.

Sec. 101: ***Requirements reforming health insurance marketplace.***

Sec. 111: ***Prohibiting preexisting condition exclusions.***

A qualified health benefits plan may not dictate exclusion of any pre-existing condition.

Sec. 112: ***Guaranteed issue and renewal for insured plans.***

The requirements of sections 2711 and 2712 guarantee availability and renewability of health insurance coverage, shall apply to individuals and employers in all individual and group health insurance coverage, whether offered to individuals or employers through the Health Insurance Exchange.

Sec. 113: ***Insurance rating rules.***

The premium rate charged, under the public health insurance options, for an insured qualified health benefits plan and coverage, may not vary— with some exceptions.

Sec. 114: ***Nondiscrimination in benefits; parity in mental health and substance abuse disorder benefits.***

A qualified health benefits plan shall comply with standards established (by the Commissioner) to prohibit discrimination in health benefits or benefit structures.

Sec. 115:   ***Ensuring the adequacy of provider networks.***

A qualified health benefits plan that uses a provider network for items and services shall meet such standards respecting provider networks as the Commissioner may establish to assure the adequacy of such networks.

Sec. 116:   ***Ensuring value and lower premiums.***

A qualified health benefits plan shall meet a medical loss ratio as defined by the Commissioner.

Sec. 121:   ***Coverage of essential benefits package.***

A qualified health benefits plan shall provide coverage that meets the benefit standards adopted under section 124 for the essential benefits package

Sec. 122:   ***Essential benefits package defined.***

The term essential benefits package signifies health benefits coverage, consistent with standards adopted.

Sec. 123:   ***Health Benefits Advisory Committee.***

Established is a private-public advisory committee; a panel of medical and other experts known as—the Health Benefits Advisory Committee—to recommend covered benefits and essential, enhanced, and premium plans.

Sec. 124:   ***Process for adoption of recommendations; adoption of benefit standards.***

Sec. 125:   ***Prohibition of discrimination in healthcare services based on religious or spiritual content.***

Sec. 131:   ***Requiring fair marketing practices by health insurers.***

Sec. 132:   ***Requiring fair grievance and appeals mechanisms.***

Sec. 133: ***Requiring information transparency and plan disclosure.***

A qualified health benefits plan shall comply with standards established by the Commissioner to accurately and timely disclose plan documents, plan terms and conditions, claims payment policies, etcetera.

Sec. 134: ***Application to qualified health benefits plans not offered through the Health Insurance Exchange.***

A Qualified Health Benefit Plan (QBHP) shall comply with the requirements of Social Security Act in regard to a qualified health benefits plan that offers in the same manner a Medicare Advantage organization.

Sec. 138: ***Information on end-of-life planning.***

The QHBP shall distribute information related to end-of-life planning; shall offer the option to establish advanced directives and physician's orders for life sustaining treatment according to the laws of the State, and information related to other planning tools

Sec. 139: ***Utilization review activities.***

Each qualified health benefit plan, and each QHBP offering entity offering such plan—shall provide adequate notice in writing to any participant or beneficiary under such plan, whose claim for benefits has been denied.

A QHBP entity shall provide for an external appeals process that meets the requirements set forth.

Sec. 141: ***Health Choices Administration; Health Choices Commissioner.***

An independent agency in the executive branch of the Government shall be established—Health Choices Administration and Commissioner

Sec. 142: **Duties and authority of Commissioner.**

Sec. 152: **Prohibiting discrimination in healthcare.**

All healthcare and related services shall be provided with no consideration to personal characteristics extraneous to the provision of high quality healthcare or related services.

Sec. 153: **Whistleblower protection.**

Sec. 156: **Application of State and Federal laws regarding abortion.**

Nothing in this Act shall be construed to have any effect on State laws regarding the prohibition of coverage, funding, or requirements on abortions, including parental notification or consent for the performance of an abortion on a minor.

Sec. 157: **Non-discrimination on abortion and respect for rights of conscience.**

A Federal agency and any State or local government that receives Federal financial assistance under this Act, may not subject any individual or healthcare entity to discriminate, or to require any health plan created or regulated, to subject any individual or healthcare entity to discriminate, on the basis that the healthcare entity does not provide, pay for, provide coverage of, or refer for abortions.

Sec. 161: **Ensuring value and lower premiums.**

Each health insurance issuer that offers health insurance coverage in the general/public market shall provide that, for any plan year in which the coverage has a medical loss ratio below a level specified by the Secretary, the issuer shall provide, as specified by the Secretary—for rebates to enrollees of payment sufficient to meet such loss ratio.

Sec. 163: **Ending health insurance denials and delays of necessary treatment for children with deformities.**

Adding at the end the following new section amends title XXVII of the Public Health Service Act: SEC. 2708. STANDARDS RELATING TO BENEFITS FOR MINOR CHILD'S CONGENITAL OR DEVELO PMENTAL DEFORMITY OR DISORDER.

Sec. 164: **Administrative simplification.**

Part C of title XI of the Social Security Act is amended by inserting after section 1173 the following new sections: SEC. 1173A. STANDARDIZE ELECTRONIC ADMINISTRATIVE TRANSACTIONS.

Sec. 165: **Expansion of electronic transactions in Medicare.**

Sec. 166: **Reinsurance program for retirees.**

Not later than 90 days after the date of the enactment of this Act, the Secretary of Health and Human Services shall establish a reinsurance program to provide reimbursement, to assist participating employment-based plans with the cost of providing health benefits to retirees and to eligible spouses, surviving spouses and dependents of such retirees.

Sec. 167: **Limitations on pre-existing condition exclusions in group health plans and health insurance coverage in the group and individual markets in advance of applicability of new prohibition of preexisting condition exclusions.**

A health insurance issuer that provides individual health insurance coverage may not impose exclusion of pre-existing condition.

Sec. 201: ***Establishment of Health Insurance Exchange; outline of duties; definitions.***

Established within the Health Choices Administration and under the direction of the Commissioner a Health Insurance Exchange, in order to facilitate access of individuals and employers, through a transparent process, to a variety of choices of affordable, quality health insurance coverage, including a public health insurance option; the Commissioner shall accept bids from, and negotiate and enter into contracts with QHBP offering entities for health benefits plans through the Health Insurance Exchange; facilitate outreach and enrollment in such plans of Exchange; to be conduct such activities related to the Health Insurance Exchange as required, including the establishment of a risk pooling mechanism.

Sec. 202: ***Exchange-eligible individuals and employers.***

All individuals are eligible to obtain coverage through enrollment in an Exchange-participating health benefits plan offered through the Health Insurance Exchange, unless such individuals are enrolled in another qualified health benefits plan or other acceptable coverage.

Sec. 203: ***Benefits package levels.***

The Commissioner shall specify the benefits made available under Exchange-participating health benefits plans during each plan year.

Sec. 207: ***Health Insurance Exchange Trust Fund.***

Created within the Treasury of the United States a trust fund known as the Health Insurance Exchange Trust Fund consisting of such amounts as may be appropriated or credited to the Trust Fund under this section or any other provision of law. The Commissioner shall pay from the Trust Fund amounts as the Commissioner determines are necessary to make payments to operate the Health Insurance Exchange.

Sec. 208: ***Optional operation of State-based health insurance exchanges.***

A State (or group of States) applies to the Commissioner for approval of a State-based Health Insurance Exchange to operate in the State (or group of States), and Commissioner shall approve a State-based Health Insurance Exchange if it meets the requirements for approval under subsection.

Sec. 209: ***Limitation on premium increases under Exchange-participating health benefits plans.***

The annual increase in the premiums charged under any Exchange-participating health benefits plan may not exceed 150 percent of the annual percentage increase in medical inflation for the 12-month period ending in June.

Sec. 221: ***Establishment and administration of a public health insurance option as an Exchange-qualified health benefits plan.***

The Secretary of Health and Human Services shall provide Exchange-participating health benefits plan that ensures choice, competition, and stability of affordable, high quality coverage throughout the United States.

Sec. 222: ***Premiums and financing.***

The Secretary shall establish geographically-adjusted premium rates for the public health insurance option in a manner that complies with the premium rules established by the Commissioner for Exchange-participating health benefit plans; and at a level fully sufficient to finance the costs.

Sec. 223: ***Negotiated payment rates for items and services.***

Secretary shall negotiate payment rates for the public health insurance option for services and healthcare providers consistent with this section.

Sec. 224: ***Modernized payment initiatives and delivery system reform.***

The Secretary may utilize innovative payment mechanisms and policies to determine payments for items and services under the public health insurance option. The payment mechanisms and policies may include patient-centered medical home and other care management payments, accountable care organizations, value-based purchasing, and bundling of services, differential payment rates, performance or utilization based payments, partial capitation, and direct contracting with providers.

Sec. 225: ***Provider participation.***

The Secretary shall establish conditions of participation for healthcare providers under the public health insurance option, i.e., Licensure or Certification, Payment Terms for Providers, Exclusion of Certain Providers.

Sec. 226: ***Application of fraud and abuse provisions.***

Provisions of law identified by the Secretary that impose sanctions with respect to waste, fraud, and abuse under Medicare shall also apply to the public health insurance option.

Sec. 227: ***Application of HIPAA insurance requirements.***

The requirements of sections 2701 through 2792 of the Public Health Service Act shall apply to the public health insurance option in the same manner as they apply to health insurance coverage offered by a health insurance issuer in the individual market.

Sec. 228: **Application of health information privacy, security, and electronic transaction requirements.**

Part C of title XI of the Social Security Act, relating to standards for protections against the wrongful disclosure of individually identifiable health information, health information security, and the electronic exchange of healthcare information, shall apply to the public health insurance option in the same manner as such part applies to other health.

Sec. 229: **Enrollment in public health insurance option is voluntary.**

Nothing in this division shall be construed as requiring anyone to enroll in the public health insurance option. Enrollment in such option is voluntary.

Sec. 241: **Availability through Health Insurance Exchange.**

In the case of an affordable credit eligible individual enrolled in an Exchange-participating health benefits plan—the individual shall be eligible for an affordability premium credit to be applied against the premium for the Exchange-participating health benefits plan in which the individual is enrolled.

Sec. 242: **Affordable credit eligible individual.**

Sec. 243: **Affordable premium credit.**

The affordability premium credit is in an amount equal to the amount by which the premium for the plan exceeds the affordable premium amount specified for the individual.

Sec. 245: **Income determinations.**

The affordability premium credit is in an amount equal to the amount by which the premium for the plan exceeds the affordable premium amount specified for the individual.

Sec. 246: *No Federal payment for undocumented aliens.*

Nothing shall allow Federal payments for affordability credits on behalf of individuals who are not lawfully present in the United States.

Sec. 251: *Establishment.*

The Commissioner shall establish a Consumer Operated and Oriented Plan program under which the Commissioner may make grants and loans for the establishment and initial.

Sec. 252: *Start-up and solvency grants and loans.*

The Commissioner, acting through the CO-OP program, may make— loans to cooperatives to assist such cooperatives with start-up costs, and grants to cooperatives to assist such cooperatives in meeting State solvency requirements in the States in which such cooperative offers or issues insurance coverage.

Sec. 301: *Individual responsibility.*

An individual's responsibility to obtain acceptable coverage, is in section 59B of the Internal Revenue Code of 1986 (as added by section 401 of this Act).

Sec. 311: *Health coverage participation requirements.*

An employer meets the requirements of this section if such employer does all of the following: (1) the employer offers each employee individual and family coverage under a qualified health benefits plan or under a current employment-based health plan. (2) If an employee accepts such offer of coverage, the employer makes timely contributions towards such coverage in accordance with section 312. (3) If an employee declines such offer but otherwise obtains coverage in an Exchange-participating health benefits plan the employer shall make a timely contribution to the Health Insurance Exchange with respect to each such employee in accordance with section 313.

Sec. 312: **Employer responsibility to contribute towards employee and dependent coverage.**

Sec. 313: **Employer contributions in lieu of coverage.**

A contribution is made to an employee, if such contribution is equal 8 percent of the average wages paid by the employer, during the period of enrollment (determined by taking into account all employees of the employer and as the Commissioner provides, including rules providing for the appropriate aggregation of related employers).

Sec. 315: **Regionalized communication systems for emergency response.**

The Secretary, acting through the Assistant Secretary for Preparedness and Response, shall award not fewer than 4 multiyear contracts or competitive grants to eligible entities; to support demonstration programs that design, implement, and evaluate innovative models of regionalized, comprehensive, and accountable emergency care systems.

Sec. 321: **Satisfaction of health coverage participation requirements under the Employee Retirement Income Security Act of 1974.**

Sec. 322: **Satisfaction of health coverage participation requirements under the Internal Revenue Code of 1986.**

Sec. 323: **Satisfaction of health coverage participation requirements under the Public Health Service Act.**

Sec. 401: ***Tax on individuals without acceptable healthcare coverage & Sec. 59B. Tax of individuals without acceptable healthcare coverage.***

In the case of any individual who does not meet the requirements at any time during the taxable year, there is imposed a tax equal to 2.5 percent of (1) the taxpayer's modified adjusted gross income for the taxable year, over (2) the amount of gross income specified with respect to the taxpayer.

Sec. 412: ***Responsibilities of non-electing employers.***

There is imposed on every non-electing employer an excise tax, equal to 8 percent of the wages paid by him with respect to employment. In cases of small employers, the applicable percentage determined in accordance with the table provided for 8 percent shall apply.

Sec. 441: ***Surcharge on high income individuals & Sec. 59C. Surcharge on high income individuals.***

In the case of a taxpayer other than a corporation, there is imposed a tax equal to (1) 1 percent of so much of the modified adjusted gross income of the taxpayer as exceeds $350,000 but does not exceed $500,000; (2) 1.5 percent of so much of the modified adjusted gross income of the taxpayer as exceeds $500,000 but does not exceed $1,000,000; and (3) 5.4 percent of so much of the modified adjusted gross income of the taxpayer as exceeds $1,000,000.

In the case of any taxpayer making a joint return under section 6013 or a surviving spouse, subsection (a) shall be applied by substituting for each of the dollar amounts therein equal to (1) 50 percent of the dollar amount so in effect in the case of a married individual filing a separate return, and (2) 80 percent of the dollar amount so in effect in any other case.

Sec. 510: **Standards for Accessibility of Medical Diagnostic Equipment.**

Sec. 715: **Protection against post-retirement reduction of retiree health benefits.**

Every group health plan shall contain a provision which expressly bars the plan, or any fiduciary of the plan, from reducing the benefits provided under the plan to a retired participant, or beneficiary of such participant if such reduction affects the benefits provided to the participant or beneficiary as of the date the participant retired for purposes of the plan and such reduction occurs after the participant's retirement unless such reduction is also made with respect to active participants.

Sec. 748: **Training of Medical Residents in Community-Based settings.**

The Secretary shall establish a program for the training of medical residents in community-based settings consisting of awarding grants and contracts under this section.

Sec. 765: **Enhancing the Public Health Workforce.**

The Secretary, acting through the Administrator of the Health Resources and Services Administration, and in consultation with the Director of the Centers for Disease Control and Prevention, shall establish a public health workforce training and enhancement program consisting of awarding grants and contracts.

Sec. 801: **Election of an employer to be subject to national health coverage participation requirements.**

An employer may make an election with the Secretary to be subject to the health coverage participation requirements.

An election may be made at such time and in such form and manner as the Secretary may prescribe.

Sec. 802: *Treatment of coverage resulting from election.*

Sec. 803: *Health coverage participation requirements.*

Sec. 804: *Rules for applying requirements.*

In the case of any employer, which is part of a group of employers, who are treated as a single employer, the election under section 801 shall be made by such employer as the Secretary may provide. Any such election, once made, shall apply to all members of such a group.

Sec. 805: *Termination of election in cases of substantial noncompliance.*

Sec. 806: *Regulations.*

The Secretary may promulgate such regulations as may be necessary or appropriate to carry out the provisions of this part, in accordance with section 324(a) of America's Affordable Health Choices Act of 2009. The Secretary may promulgate any interim final rules as the Secretary determines are appropriate to carry out this part.

Sec. 809: *REPORTS.*

The Secretary shall submit to the Congress a separate annual report on the activities carried out under each of sections 811, 821, 836, 846A, and 861 (yet to be drafted)

Sec. 871: *FUNDING.*

For the purpose of carrying out parts B, C, and D (subject to section 845(g)), there are authorized to be appropriated such sums as may be necessary for each fiscal year through fiscal year 2014.

Sec. 872: **Funding through Public Health Investment Fund.**

For the purpose of carrying out this title, in addition to any other amounts, authorized to be appropriated for such purpose, there are authorized to be appropriated, out of any monies in the Public Health Investment Fund,

Sec. 931: **Center for Quality Improvement.**

There is established the Center for Quality Improvement to be headed by the Director. The Director shall prioritize areas for the identification, development, evaluation, and implementation of best practices for quality improvement activities in the delivery of healthcare services.

H.R. 3200 Funding

Exactly how ObamaCare is going to be funded is not quite clear yet. One thing for sure is the hidden tax increases imbedded in the bill, e.g., 3.8 percent surtax on all investment income; 3.5 percent fee for insurers in states opting-out of healthcare exchanges; 2.3 percent excised tax on sales by medical device makers; 7.5 percent to 10 percent threshold increase for medical expense income tax deduction; Medicare payroll tax hike from 2.9 percent to 3.8 percent for earnings over $250,000; 8 percent employer's penalties for each employee without approved healthcare coverage; 2.5 percent of adjusted income penalty for persons without healthcare insurance; and etcetera. CBO estimates, that with all these changes, the healthcare cost will rise to 40 percent by 2018—a rather pessimistic prediction.

The following main sources for funding ObamaCare floated around are the following.

PROPOSED CBO SPENDING CUTS from 2013 to 2022:

- $415 billion in Medicare payment rates

- $156 billion in Medicare Advantage Payment
- $56 billion cuts on Disproportionate Share Payments
- $114 billion Other Cuts
- **TOTAL Cuts: $741 billion**

PROPOSED CBO REVENUE INCREASES from 2013 to 2022:

- $55 billion penalties paid by uninsured
- $106 billion tax penalties paid by non-electing employers
- $111 billion excised tax of Cadillac Plans
- $216 billion effects of increasing coverage
- $318 billion additional Hospital Insurance Tax
- $87 billion fees on certain Manufacturers and Insurers
- **TOTAL Increases: $893 billion**

The projected total cost of OBAMACARE from 2013-2022 is $1.634 trillion or a yearly budget of $163.40 billion. Like every other government programs in the past, the ObamaCare budget will soar. Medicare and Medicaid in 1965 had a combined funding of $0.30 billion. By 1975, Medicare and Medicaid expenditure raised 69.67 times to $20.9 billion. This increase was in10 years. Using the same rate of increase, ObamaCare by 2022 is likely to cost the nation $11.384 trillion.

Incidentally Medicare and Medicaid had increased from $0.30 billion in 1965 to $787 billion in 2010—an increase of 2,623.33 times.

Concluding Remarks on ObamaCare

Let us assume that ObamaCare conservatively grows to 30 times by the year 2022. That will be $4.902 trillion. In 2010, the nation's total healthcare expenditure is $2.52 trillion will also increase 30 times to $75.6 trillion by 2022. By 2022, the total healthcare expenditure at a conservative 30 times increase is $80.50 trillion or about 5 times the national GDP.

Surprisingly many of the supporters of HR3200 while still being legislated have applied for ObamaCare waivers. As of May 13, 2011 there were 1,372 large companies and unions applying and approved to opt-out of ObamaCare; the likes of McDonald's, Waffle House, Universal Orlando, Ruby Tuesday, AMB Bowling Worldwide; and the local chapters of the International Brotherhood of Trade Unions Health and Welfare Fund and the Teamsters. Worse case scenario is what Wal-Mart (the nation's largest private employer) is planning to drop from its health benefit programs newly hired employees working fewer than 30 hours a week; qualifying them for expanded Medicaid program. Roughly 1.4 million U.S. workers are vulnerable losing medical insurance. If Wal-Mart does this, many more big employers will do the same, and that will really accelerate the cost of the ObamaCare. Lastly there are 36 States opting-out of the ObamaCare. This tells us one thing. ObamaCare is not as good as it was characterized and sold to the American people.

While ObamaCare has humanitarian aspects and looks very good on the surface, studying the individual provisions reveal a very complex healthcare bill that is predominantly federal government regulations in all aspects of the U.S. healthcare system. The worse is that this healthcare law did not address at all the various issues that had driven the present healthcare to levels almost unaffordable by average American, and will destroys our economy and our country. It failed to meet its primary purpose stated in Sec. 100.

Although the original intent of HR3200 is to provide healthcare coverage for the 47 million uninsured Americans, Patrick Burke on August 8, 2012 published a CBO report that 30 million Americans will continue to have no healthcare coverage by 2022. Insuring all Americans was the primary argument, and the promise made to pass ObamaCare.

We the people have to force our elected officials in Washington, DC to repeal ObamaCare. We must encourage them to start from scratch and to include the various principles of a free market economy that make a U.S. healthcare sustainable and affordable.

The reason for the repeal: ObamaCare failed its principal objective to provide healthcare coverage for all Americans without exception at an affordable and sustainable price.

V

A Review of Established National Health Systems

How do other healthcare systems in the world compare with America's? American healthcare is not perfect. But neither are the other healthcare systems. Though critics of American healthcare strongly endorse socialized health systems like those of Canada, England, France, Germany, Japan, Sweden, and Switzerland, many faults can be easily identified in these other healthcare schemes. Yet, the majority of the people in those countries are satisfied with what they have. The success or failure of a country's medical care system therefore boils down in part to the preference, tolerance, acceptance, and the ability of the citizens to make it work. What is perceived as good in another country's healthcare much smaller than the United States will probably not work in this country simply because the U.S. has a much larger population from different cultural backgrounds and a huge geographical area that extends four time zones.

As an American, would you like to have a healthcare system similar to any of the countries reviewed? You be the judge.

The British National Health Services

The *National Health Services (NHS)* of England was established in 1948. Its main purpose is to provide free healthcare *"at-point-of-need"* to all the citizens of the United Kingdom. It is a socialized healthcare system in its purest form. This means free doctor's visits and hospital treatments.

111

The National Health Services has a Minister or Secretary of Health directly accountable to England's Parliament. She/he is responsible for the performance of the Department of Health that carries out the different functions of the NHS. The Department of Health is in charge of regulating, planning, and inspecting healthcare services around the country, and preparing policies for the various branches of the NHS. It has two main divisions: Primary and Secondary Healthcare Services or better known as Primary and Secondary Care Trusts.

The NHS is run by layer upon layer of bureaucracy called Trusts both in the central and local government. Locally there are 28 strategic healthcare authorities or commissioners, and 300 Primary Care Trusts in England. The numerous health authorities at various levels appear to hamper the NHS strategy for efficient healthcare delivery to the citizens of England. There is too much bureaucratic red tape with too much regulation and oversight at every level of the healthcare hierarchy.

The Primary Care Trusts' main function is to determine the local healthcare needs, and make sure there are enough primary healthcare providers in the area. These are the general/primary physicians, dentists, optometrists, pharmacists, etcetera. Receiving 75 percent of the NHS budget, the Primary Care Trusts basically control and dictate who are referred to the Secondary Care Trusts. They determine the extent and quality of services given by the hospitals, physician specialists, dentists, patient transportations, and other population healthcare initiatives. Lately, outsourcing of patients is done to private clinics and other healthcare entities for after-office-hour medical coverage. Outsourcing of patient care is also made to reduce waiting list of patients scheduled for surgeries, specialized therapy, diagnostic procedures, etcetera.

The Secondary Care Trusts' primary function is to provide medical services to patients requiring specialist, patient transport, and hospital confinement. It is administered by different trusts.

- Acute trusts supervise hospitals providing short-term care such as accidents and emergency, maternity, surgery, x-ray, and other diagnostic and treatment procedures.

- Care trusts carry out a variety of services both in medical, mental, and social work, and are generally setup to meet and work closely with NHS and local health authorities.

- Mental health trusts are instrumental in providing psychiatrist, psychologist, psychotherapist, and psychotherapy. They also provide the training grounds for future mental health providers.

- Ambulance trusts provide needed transportation for patients. There are thirty ambulance services in England; each is operated by its own trust.

Foundation Trusts are subsets of the Secondary Care Trusts. These are hospitals that have been designated and determined to be highly efficient and effective, and usually owned by the local communities. The local residents have the prerogative to manage their hospital with less interference from central supervision and monitoring. As a result, these foundation hospitals are financially more flexible and better run. They are controversial.

Generally, the government through taxation finances 82 percent of healthcare funding. Other sources of funding are 13 percent from employer/employee contributions, and 4 percent from user fees. The private healthcare sector is much smaller than the NHS. It had no place in the NHS prior to Margaret Thatcher's Health Privatization Initiative. Now about 10 percent of the English population participates in private healthcare to avoid unduly long delays, and to receive timely medical care, while 90 percent of the populations are still enrolled with the NHS.

Private hospital are owned and operated by private hospital groups normally licensed by the local health authorities where the hospital is located. They are inspected twice a year but are not regulated by the national health inspection authorities that monitor NHS hospital facilities. These private institutions are frequently used to make available medical specialists, diagnostic procedures, surgical treatments, and other health associated needs such as cosmetic surgeries, and treatment and rehabilitation of addicts. Private medical care is largely through private health insurances. It mirrors the NHS but it is not required to follow national treatment guidelines and other healthcare policies. It does not have any direct responsibilities to the wider communities.

The NHS is a gatekeeper healthcare system. It means that patients have to see first a general practitioner or primary care MD before they are referred to specialists or specialized treatment and diagnostic centers. Physician's compensation is dependent on capitation. A small amount of money is paid upfront for a patient by the NHS either on a monthly or yearly basis. From the patients' capitation money is taken the medical expenses of the patient who needs care. The longer the list of patients a doctor has, and the larger the number of relatively healthy patients she/he has, the bigger is the compensation if few uses it with minor medical care. The downside is a few very sick patients can easily wipeout the collected capitations; subjecting the physician financially at risk. Because a patient chooses her/his primary physician, and a physician can reject a patient by simply saying her/his list is full, a doctor can easily select her/his patient population of healthy people. Doctors, who have healthy patient loads, do well in the system, while the ones with a sick patient population are financially compromised, some file bankruptcy. Underutilization by medical practitioners of the health system then becomes understandable. Many in the healthcare profession left the NHS system and migrated to other European countries and the United States because of uncertain economic return in the system.

The considerable loss of health professions in the country prompted the government to increase the compensation of the English health professionals and later allowed them to incorporate private patients in their practices. The act added to the already high health expenditure, and sent the national healthcare budget soaring, taking 40 percent of the government's discretionary spending in 2002, representing 7.6 percent of gross domestic product. Even if expenses in other areas are restricted at 3 percent increase, a 7 percent rise for the NHS spending will still take in excess of 60 percent of all available extra cash the British government has by the year 2008.

England's NHS is facing a massive financial burden with its aging population like those in other European countries, Japan, and the United States. With a low gross domestic product it will be hard for the English government to find resources to finance the growing demand for healthcare. Underfunding of the system will further aggravate existing problems of healthcare providers' shortage, long delays, limited use, and availability of modern technologically advance medical equipment, and the rationing of expensive medical and surgical therapy. Underpaid, overworked, and stressed-out healthcare providers eventually lead to unresponsive, disrespectful, uncaring, and unfriendly attitudes and animosity between providers and patients.

Let's examine the delays in the NHS. In an article written by Conrad F. Meire he observed, "All too frequently they (patients) don't get care at all; are subjected to queuing for twelve months or more; get better on their own; are sent to other countries for care; are shifted to the private sector for care; or die while on the waiting list to see a doctor or gain access to a hospital." In 2000 there were 1,120,000 patients who waited uncomplainingly for necessary in-hospital treatments; 40,000 patients were waiting over twelve months for surgery. And because of underfunding, elective but necessary and expensive medical and surgical treatments such as coronary by-pass, hip or knee replacement, renal dialysis, etcetera, are often not done. Compared with the U.S., U.K. performed 20 percent or less of these elective treatments. Another criticism for the British NHS is its single-payer health system, which Mr. Meire commented, "… is a cruel joke being played on gullible citizens. If not a cruel joke, then it is at best a hollow political promise to treat everyone the same while consistently breaking faith and treating no one the same."

How do the English citizens rate their healthcare system? Polls showed 48.1 percent satisfied, 11 percent neutral and 40.9 percent dissatisfied, 14.6 percent think it is well run, 27.4 percent think it needs minor changes, 42 percent think it needs fundamental changes, 14 percent think it needs to be rebuilt, and 2 percent have no opinion.

With huge increases at the various levels of the NHS, the British government is finding it harder and harder to find extra money to finance its healthcare system in its present form.

The Canadian Medicare

The Canada Health Act created Canadian Medicare is a socialized healthcare system consisting of a group of national health insurance plans. Although the health plan is primarily financed by the federal government to meet the needs of the citizens, it is implemented at the provincial and territorial levels of government. William Mackenzie King introduced the concept of universal health insurance in 1919. This initial concept of a national healthcare plan commenced in 1947 when Saskatchewan passed the first universal hospital insurance plan sponsored by the provincial government.

By 1961 the other provinces and two territories joined Saskatchewan's lead when laws were passed to allow the federal government of Canada to share the cost of the provincial hospital insurance plans. The universal provincial hospital insurance program was limited to inpatient hospital coverage. In1968 and 1972 the federal government of Canada instituted corrective measures, and in 1984 the Canada Health Act made Canada's healthcare into a comprehensive medical care plan in its present form—referred to as *Canada's Medicare*. Medical services included both outpatient and in-hospital treatment, and a variety of other medical benefits. Not included in the system were many other healthcare needs such as dental care, optometric services, prescription drugs/medications, etcetera.

Canada's Medicare was founded on the following principles to achieve Medicare's main objectives. These were to be carried out primarily by public health agencies. There were no provisions included in the Canada Health Act for any possible role private clinics, hospitals, and/or private health insurances may play. Nevertheless, in years to come there existed private medical clinics, hospitals, and insurances that supplemented Canadian healthcare. They were tolerated and ignored, and are now actively, politically, and publicly debated as to the possible important function of private healthcare providers.

- Public administration means that Canada's Medicare must be administered by the provincial health insurance public authority, and be accountable for its proper and cost-effective operation as a not-for-profit government entity.

- Comprehensiveness avails provincial citizens a full range of basic medical care.

- Universality entitles every covered member the same level of healthcare.

- Portability is the right of a provincially insured person to be treated outside her/his province or territory of residence for a specified period of time.

- Accessibility is the privilege of an insured person to be seen by healthcare providers and health facilities.

Access to Canada's Medicare requires the registration of each citizen to the provincial or territorial health department with the exceptions of the inmates, the Canadian Armed Forces, and certain members of the Royal Canadian Mounted Police who are provided with a different healthcare plan. The registration form contains basic personal information for obvious purposes of recording medical information and billing. After enrolling in the provincial health insurance plan, the enrollee is given a health card and is instructed to signup with a healthcare provider and is somewhat "locked-in" with that provider. The health card contains a number that tracks the patient's medical activities. A person has only to present her/his health card to see the provider. No paperwork is necessary or done.

Funding of Canadian Medicare is by both the federal and provincial/territorial governments from business and personal taxes. In some provinces or territories additional sources of funds are in sales tax and lottery proceeds. Provinces like Alberta, British Columbia, and the city of Toronto also collect an additional premium to help fund the provincial healthcare insurance. Generally the federal and provincial governments contribute approximately 75 percent of the provincial health budget and 25 percent from private businesses and insurances.

Canada's healthcare like every industrialized country has a significant impact in the economy. In 2001 Canada's healthcare spending reached $100 billion. From 2002 to 2004 the healthcare expenditure grew at an average annual rate of 7 percent to $130,300,000,000; approximately 9.5 percent of Canada's Gross Domestic Product, compared with the U.S. of 14.6 percent. Though healthcare expenses appear under control, the reality is that the federal government has been adding more and more money into the health system. The expansion of healthcare benefits and expenditures has been a big political debate between the liberal and conservative political parties. The 2002 Romanow Commission on the Future of Healthcare in Canada recommended another $41,000,000,000 infusion into the healthcare to correct and upgrade deficiencies in Canada's Medicare. The commission's recommendations have been approved lately. The areas of great concern are for Rural and Remote Access, Diagnostic Services, Primary Healthcare, Home Care, and Catastrophic Drug. What all these mean is to address issues such as reasonably easy access in rural and remote areas, timely diagnostic service and treatment, removal of obstacles in innovative primary care delivery, development of a national home care strategy, and providing Canadians for more expensive drug coverage. Included in the legislation is the formation of a *National Electronic Medical Record keeping with adequate protection to individual privacy by amending the Criminal Code of Canada.* Undoubtedly all these new programs will add significant expenses to Canada's Medicare.

There seems some ambivalence in regards to private insurance. Some provinces want private insurance to play much greater roles, and others don't want private insurances to have any role. The most common argument for the pros is that private insurances can supplement the mounting cost of healthcare; those for the cons, it creates a two-tier healthcare system. Actually there is great disagreement going on among politicians in regard to the integration of private insurance into the Canadian healthcare industry. It is politically a hot topic. Nonetheless, employers frequently offer private healthcare insurance as an employment benefit package. Some Canadians also buy private health insurance plans as a supplement to provide health insurance for services not covered by Medicare, and avoidance of long-waiting list in the public healthcare sector.

Although the Canadian Medicare is a socialized publicly financed health program, the majority of healthcare providers in the system are private medical practitioners. Private hospitals and clinics are not legally sanctioned but tolerated to operate because they give capably better and faster healthcare services with much less delay. They are reimbursed 80 percent of their fees by private insurances.

Healthcare providers—primary care and specialist physicians, oral surgeons, and public hospitals—are covered by the Canada Health Act. Compensation is fee-for-service controlled by the government. Professional fees are significantly lower compared to those of the U.S. representing approximately 30 percent of a typical American usual and customary fee. Many believed this led to many healthcare professionals migration to the south leading in shortages of physicians, nurses, and other highly qualified healthcare providers.

The Canadian Medicare is regarded as one of best by virtue of its universal coverage, low infant mortality rate of 6.3 per one thousand live births compared with the U.S. of 8.3 per one thousand live births, longer life expectancy of 81 years for women and 74.5 for men versus 78.9 years for women and 72.1 years for men in the U.S., and relatively cheaper healthcare than the United States. While 80 percent of Canadians are satisfied with their access to the healthcare system, there is increasing discontent and frustration; some strongly recommend restructuring the healthcare system.

A Canadian Medical Association poll in 2004 revealed 14 percent of Canadians think they have sufficient number of healthcare professionals and yet 49 percent expressed longer waits than anticipated to see a medical specialist or their family doctors. Seventy-five percent of Canadians complain of long delays to get necessary emergency room, diagnostic procedure, and treatment services. There is significant concern on medical conditions deteriorating due to untimely diagnosis and treatment. The following examples should illustrate how much delays are encountered by Canadians seeking healthcare services: referral to a general/primary care practitioner, 17.7 weeks vs. same day to two days in America; referral to medical oncologist, 6.1 weeks in Canada vs. few days to one week in the U.S. and radiation therapy for cancer, 8.1 weeks in Canada vs. one week in the U.S.; routine hip replacement, six to12 months for Canadians vs. one to three weeks for Americans; and for diagnostic imaging such as CT and MRI scans, two to three months in Canada vs. a few days in America.

Because of long waiting list to get needed diagnostic tests, and the unnecessary delays in receiving expensive but appropriate specialized treatments like elective surgical procedures for knee replacements, cardiovascular by-pass, and other life-threatening but non-emergency conditions and huge bills for prescription drugs, dental and other health services not covered by the Canadian Medicare, disappointment with the healthcare system is growing faster than anticipated.

Though rated one of the best national healthcare plans in the world, Canadian Medicare still has legitimate problems that perhaps will take years if not decades to overcome and correct. The Canadian federal government spends 22 percent of its tax revenues on healthcare. That is an enormous amount of money Canadians spend on healthcare services they are not assured of getting on a timely basis.

The French National Health Insurance System

The French *National Health Insurance (NHI)* system has its beginning in 1930 when the French government passed into law compulsory health insurance coverage to all employees. This was extended to include farmers in 1961 and the self-employed in 1966. The French social health insurance is the blending of two ideologies: that of liberalism and egalitarianism. It is a unique compromise of the public and private; it is complicated with an array of healthcare providers and payers, each one having some degree of independence from the government.

The French government has a heavy-handed style of managing the NHI. Before the 1996 Juppe Reform, central government tightly controlled the training of healthcare personnel, quality of health services, volume of healthcare supplies, work conditions, and social programs. It dictated methods of financing, setting tariffs or provider's fee, population coverage, and regulated healthcare diagnostic and treatment procedures. With the Juppe Reform, the French Parliament was given the ultimate authority for setting up objectives and the yearly budget of the social security program. Three committees were established; the High Committee of Public Health, the National Health Conference, and the Conferences Regionales de Sante.

The High Committee of Public Health—chaired by the health minister—determined and identified the health objectives of the nation. Under the Ministry of Health are two large organizations: the General Health Management and Hospital, and the Healthcare Management. The National Health Conferences made of representatives from various healthcare agencies analyzed and determined healthcare priorities. The Conferences Regionales de Sante studied local health needs and recommended public health priorities. The disbursement of funds is under the authorities of the Ministry of Finance, and Ministry of Social Affairs and Employment. There are also 22 regional health and social work bureaus whose primary functions are to plan for local health and social affairs, which include setting and determining budget limits, number of hospital beds, instillation of expensive medical equipment and treatment facilities, and other health-related funds. In addition there are the Caisse Nationale d'Assurance Maladie des Travailleurs Salaries (CNAMTS), Caisse Protectiones d'Assurances Maladies (CPAM), Caisses Regionale d'Assurance Maladie (CRAM), and the Unions Regionales des Caisse d'Assurances Maladie (URCAM), just to name a few more government regulators.

The CNAMTS is the insurance healthcare plan for employees and their families in the commercial and industrial sectors, farmers, and other professionals. It manages the employees' healthcare plan nationally through 16 regional and 133 local insurance funds. The 133 local insurance plans are responsible for the general development of local healthcare resources while the sixteen regional funds co-ordinate capital development in the region. Approximately 80 percent of the population is covered by CNAMTS. CPAM is in charge of the registration of covered members, reimbursement of claims and benefits, and other healthcare initiatives in sanitation, social works, and preventive programs in the local areas. It is fondly called "secu." The CRAM is the counterpart of CPAM in workman's compensation, workplace safety, and other health functions it is responsible for in the commercial and industrial sectors.

Some other government programs for healthcare benefits cover the other 20 percent of the population. Poor citizens and those who never worked are granted government healthcare entitlements through mechanisms similar to U.S. Medicaid. Foreigners are required generally to have their own health insurance. Special arrangement is extended to all European Union citizens since the adoption of the **European Smart Health Card** by other EU nations in 2004.

The NHI assures all French citizens access to a universal healthcare service. It is one of the four branches of the French Social Security system consisting of Health, Life, and Disability Insurance Benefits; Workman's Compensation and Work Disability Insurance Benefits; Old Age Insurance Benefits; and Family Insurance Benefits. It is funded through various French governmental schemes of compulsory employer and employee contributions, taxations, pensions, and other forms of revenues. It would appear that the NHI finances health insurance coverage for the workers using the obligatory employer and employee contributions. The healthcare insurers are non-governmental for-profit or not-for-profit companies. Insurance premium charged is based on a fix percentage of the insured's payroll amounting to 20 percent. The insurers' allegiances are to the employers and employees rather than the government.

Specifically the NHI is subsidized mainly by the National Social Security program contributing up to 75.5 percent of expenses. Supplementary sources are from private health insurances (12.1 percent), payments by patients (11.3 percent), and 1.1 percent from other general taxations. There is now a trend for the French government to broaden the tax base for the national Social Security system to lessen the dependence of NHI to payroll contributions. The French healthcare budget represents about 10.4 percent of France's GDP in 2002 compared to 14.6 percent for the United States of America. To cope with the mounting financial pressure of healthcare, expanding the role of supplementary insurance is seriously being considered. Supplemental insurances are usually tailored to some professions, and employers' insurance plans. There are other commercially available insurances the general public can purchase for personal healthcare coverage that best fit their medical requirements. There is clear indication of acceptance by the French people of a supplementary healthcare insurance to avoid rationing, long delays in public medical facilities, and to recover medical charges incurred at point-of-service. In 1960, 30 percent of the population had supplemental private healthcare insurance, and by 2001, 87 percent had it.

To access the national health benefit, French citizens have to register at the local CPAM/secu. Basic information for personal identification is taken. Once the registration is completed a "carte vitale" is issued. It has an insurance number to be presented every time healthcare benefit is sought. The carte vitale is French's health ID card designed for fast and efficient completion of healthcare services. It facilitates faster communication and exchange of information, between medical and other healthcare providers, important in the care of patients. And with the carte vitale, the patient can go or choose her/his own physician, clinic, and hospital. They are allowed a wide range of freedom in selecting their healthcare providers. It is not a means of payment but rather a means to ease reimbursement and associated general administration of healthcare records and services.

There are three basic philosophies the French healthcare system adheres to. They are personal payment, choice of physician, and freedom of practice. Patients normally pay their healthcare providers directly at point-of-use and are issued receipts with the amount of charges prominently displayed. This method of personal payment creates a sense of responsibility and serves as a deterrent to medical care abuse by some patients. About 85 percent of the doctor's fee and other medical charges are reimbursed by the government, and remaining 15 percent by the supplementary insurance, if patients have it. This reimbursement scheme also applies to hospital healthcare practitioners. Professional tariffs or fees are enforced, and annually reviewed and adjusted by NHI.

The French medical establishment has three types of institutions: public hospitals, private hospitals, and not-for-profit health facilities.

Only 32 percent of hospitals are considered public hospitals offering 97 percent of hospital beds for both acute and long-term confinements. Less than 1 percent is allocated for acute care. The private hospitals and not-for-profit health facilities provide 3 percent of hospital beds. These are, however, for acute short-term hospitalizations. The hospital medical, dental, and other paramedical staffs are compensated as hospital practitioners. Ninety-one percent of public hospital operational expenses are primarily backed up by a system of endowments paid for by the NHI since 1985.

The private hospitals evolved from private clinics established by surgeons and obstetricians. The not-for-profit healthcare facilities developed from denominational religious healthcare facilities. These not-for-profit healthcare institutions are funded similarly as the public hospitals. Between the public and denominational hospital, 50 percent of all surgeries and 60 percent of all cancer treatments in France are done.

Like in other socialized healthcare systems the French NHI is plagued with delays and long waiting lists in the public healthcare institutions. Supplemental health insurance carried by 87 percent of the population and the existence of private practice, private clinics, and hospital significantly ease overcrowding in public health institutions and over-burdened public healthcare providers.

French people rated their healthcare system as: 65.1 percent satisfied, 23.3 percent neutral, and 14.6 percent dissatisfied; 38.9 percent think it is well run, 30.9 percent think it needs minor changes, 26.6 percent think it needs fundamental changes, and 3.6 percent think it needs to be rebuilt completely.

As in other socialized medicine the public healthcare facilities are overloaded and invariably long delays occur. The France healthcare system, however, offers healthcare consumers alternative chooses between the public and private healthcare services without difficulty moving from one system to the other. The guaranteed access to medical care reflected by the general health of the population, the French healthcare system was rated the "best in the world" in 2000-2010 by the Economic Cooperation and Development (OECD) and the World Health Organization (WHO)

Germany's Healthcare System

In 1993 the German government enacted the ***Healthcare Structural Reform Act***, otherwise known as the *Gesundheitsstrukturgesetz*

(GSG), ending a period of healthcare benefits imposed under *Statutory Public Health Insurance Laws*. The statutory health insurance called *Gesetzliche Krankenversicherung (GKV)* gave the organizational framework for Germany's present system of healthcare insurance funds known as **"Sickness Funds."** Under the current system, the GSG defines and enforces the roles of the providers, payers, healthcare insurances, and hospitals but does not run the healthcare system.

A consortium of national and regional independent associations of payers and providers administer and manage the healthcare plan. They decide the specifics of the national health policies and negotiate among themselves how to handle and finance the healthcare in the region. This self-governing consortium of healthcare providers and payers are funded both by compulsory and voluntary insurance premiums from members. The members comprising of about 92 percent of the German population are employees earning a prescribed ceiling of about $40,000 a year in wages. Those below the ceiling (7.7 percent of the population), mostly civil servants and self-employed, have private for-profit health insurances. About 0.3 percent of the population carries no healthcare insurance at all. These are the rich who can afford healthcare and the very poor who are given healthcare coverage through social programs.

The fragmented organizational nature allowed the German healthcare system to avoid a single group in dictating the terms of delivery, reimbursement, compensation, quality of care, and other important healthcare issues the entire system may be confronted with. As a result universal healthcare coverage is recognized and honored by all the medical doctors, clinics, hospital, and health insurance companies. Patient access to medical care is fairly simple and immediate; the system has accomplished a comprehensive high quality healthcare and high degree of equity for the German citizens. Freedom to select their own healthcare provider is a feature other socialized healthcare systems do not have.

The German system offers two categories of healthcare insurance funds—the **Primary and Substitute Insurance Funds**. Determination which healthcare insurance fund employees are enrolled to is based on the type of profession or work and the amount of income. There are six types of primary insurance funds covering about 46 percent of the workforce. These six types are farther subdivided into different funds insuring specific classes of workers. The substitute insurance funds provide healthcare insurance to 34 percent of the population to mostly white- and blue-collar workers earning more than the ceiling income annually determined.

Membership to either type of insurance fund is compulsory for everybody with the exception of a small percentage of the German population already mentioned. Both the employers and employees are required to contribute equal amounts for the cost of the member employee's health insurance premium, which in 1990 averaged 12 to 13 percent of the employee's gross income. The amount of health insurance premium is determined not on one's state of health, number of family members, or marital status but rather on an employee's earning capacity.

In cases of two income families where the wife and husband work, both are compelled to pay the health insurance premiums based on her/his gross monthly or annual salary. The national and local governments pay the premiums of unemployed members. Retirees either pay their own premiums or by their pension plans. Invariably this arrangement created wide variations in the financial revenues of the different health insurance funds. To correct this inequity the **Healthcare Structural Reform Act** of 1993 mandated payment to financially troubled sickness funds by a *National Reserve Health Insurance Fund*—funded by the consortium—to compensate for the lower premiums collected by healthcare funds burdened by members with lower incomes and high medical expenditures attributed to a larger number of members with more serious medical conditions and expensive cares.

The principle of redistributing healthcare resources to cover every citizen regardless of social standing is strongly and politically endorsed. In its present funding the German system is thought of as inequitable by not allowing the rich to participate in providing healthcare for the poor. It is the working class that supports the entire healthcare system, and with ballooning healthcare expenses, the way Germany's healthcare system is funded is considered as regressive.

About 11 percent of Germans have private healthcare insurance. These private healthcare insurance companies are for-profit. Most of their subscribers are civil servant employees wanting 50 percent coverage of their medical bills not covered by their public insurance funds. Some Sickness Fund members find it necessary to get supplemental insurance for certain amenities like private rooms or freedom to select their own hospital attending physician and/or specialist. The self-employed individuals earning above the income ceiling are disqualified from becoming members of the public insurance funds and are forced to have private health insurance. Members of the public health insurance fund leaving for private insurance normally are permanently barred in returning to the public health insurance coverage. Premiums for private insurance are generally cheaper than the publicly-funded sickness funds, but their payments to providers for services are twice as much. Private insurance premium is based on the insurance buyer's age, and is set at the time the insurance is bought. It is adjusted accordingly on the overall cost of providing healthcare to members.

Healthcare providers are compensated on a fee-for-service scheme. Bills from healthcare providers are paid by the healthcare insurance funds as submitted by providers just like in the early development of the healthcare system. With rising healthcare cost the government passed the *German Cost Containment Act* of 1977 and created a body called the *Concerted Act (CA).* It is composed of the various healthcare providers, healthcare insurance funds, employers, employees union, and government representatives from many levels of government. CA functions to set physician and hospital fees and other healthcare-related expenses. Modest co-payments to prescription medications, dental, hospitalization, and other healthcare services were introduced. At first fees of healthcare providers were reduced proportional with the quarterly over budget of the healthcare funds. By 1986 physician's fees were capped, followed by regulations directed to the increasing healthcare cost. In spite of attempts by the central government to control healthcare expenditures it continued to increase representing 10.7 percent of Germany's GDP by 2001—the third highest in the world.

The Germans rate their healthcare system as: 66 percent satisfied, 23.1 percent neutral, 10.9 percent dissatisfied; 36.9 percent think it is well run, 38.5 percent think it needs minor changes, 16.7

percent think it needs fundamental changes, 2.4 percent think it needs to be rebuilt, and 5.5 percent have no opinion.

Japan's National Health Insurance Act

The national health insurance of Japan had its beginning in 1905 when the Kamegafuchi Textile Company gave some limited health benefit to its employees. Decades later more and more companies followed Kamegafuchi's example and by 1922 the first health insurance law was enacted to include industrial workers and miners. This mandated healthcare insurance coverage of employees by the employers acknowledged the important role of the government in Japan's healthcare. It was enforced in 1927. The healthcare insurance law was reformed to include farmers, fishermen, foresters, and other groups in 1938, and again revised in 1958 to integrate the remaining 30 percent of the population previously excluded. In 1961 the *National Health Insurance Act (NHIA)* was passed, forcing every city, town, or village government to provide healthcare insurance to every resident without healthcare benefit. This act completed the universal healthcare coverage of the Japanese people.

The Japanese NHIA features universal coverage, freedom of choice, and is mandated by the government. Coverage is dependent on income and ability to pay, and not on preexisting conditions or other health-associated actuarial risk factors. It is multi-payer and employment-related system with both employer and employee responsible in paying compulsory insurance premiums. Although healthcare is predominantly a fee-for-service practice by private clinics, hospitals, and other healthcare facilities, the government has strong regulatory controls on healthcare financing and health insurance operations. The delivery of healthcare, however, is under the control of the medical profession. So far the medical profession has done a good job; the citizens of Japan have not expressed any concern with healthcare rationing, or under- or overutilization of the system.

The National Health Insurance system is made up of 2,000 private insurance companies and more than 3,000 government agencies. It has two distinct groups of beneficiaries: the Employee and the Self-Employed Groups.

The Employee Groups cover 65 percent of the population. These are the employees and their immediate and elderly dependents. The Self-Employed Groups are the self-employed and their immediate and elderly dependents, and the unemployed with their dependents—35 percent of the population. The employee groups are farther subdivided into four different subgroups.

1) The Government Managed Insurance Plans provide healthcare coverage for the employees of companies with five to 300 employees; covering 30 percent of the population. The employer and employee equally contribute the healthcare insurance premium, fixed by law at 8.2 percent of gross employee's income.

2) The Society Managed Insurance Plans are for companies with more than 300 employees. They cover about 26 percent of the population. Payroll contribution is 5.8 percent to 9.5 percent of the employee's gross income. The employer is required by law to put in at least 50 percent of the premium and in some cases up to 80 percent. Representatives jointly administer these plans from management and labor union.

3) The Mutual Aid Association of Healthcare Insurers covers 10 percent of the population. They are the government employees. There are 27 plans for the national government employees and 54 plans for local government and quasi-public employees. Insurance premium contribution is 8.5 percent of the employee's gross salary paid for by both the employer and employee.

4) Day Laborer Plans are for individuals who work less than two months a year and seaman. They cover 0.1 percent to 0.4 percent of the population.

The self-employed groups are the employees of companies with fewer than five employees, self-employed persons and retirees on pension. They are given healthcare coverage by 3,000 municipal and national government entities, administered by 166 national health insurance societies. Premium contribution is based on reported income, financial assets, and the number of persons per household.

Japanese healthcare system is financed by compulsory tax contributions from both employers and employees ranging from 5.8 percent to 9.5 percent of the employee's gross monthly income and employee co-payments, along with some government subsidies that fund mostly elderly and the poor. The tax contributions, co-payments, and government subsidies are used for the purchase of private health insurance coverage available from the various national health insurance plans. To access healthcare benefits co-payments are required. The amount of the co-pays depends on which health insurance group employees belong. For the Employee Groups the co-pays usually are 30 percent for outpatient, 20 percent for in-hospital and 10 percent for other health-related services. For the Self-Employed Groups their co-payments are 30 percent for outpatient, in-hospital, and other health-related services. Retired individuals have co-pays of 20 percent for in-hospital care and 30 percent for outpatient care. The basis for the co-pays percentage rate is not clear. It could be percent of monthly income or percent of healthcare services incurred. Under catastrophic health plans there is a set monthly ceiling for co-payments beneficiaries are responsible for.

Though there appears some type of financial inequity among the different groups, statistical analysis showed that for every dollar paid in premium, the society managed funds receive $0.62 in healthcare reimbursements; government managed plans, $0.84; self-employed, $1.66 and retirees, $4.45.

These insurance programs and societies administer the health funds in a quasi-public status with mandated responsibilities to cover all eligible members with uniform delivery of health benefits. Healthcare benefits provide basic medical care which include outpatient, in-hospital and extended cares, some dental, and prescription drugs. Not included are abortion, cosmetic surgery, traditional medicine, hospital amenities, some highly technical diagnostic and therapeutic procedures, and childbirth.

An interesting aspect of the Japanese health insurance employee group programs is the cash benefits members are entitled to. These cash benefits are given in cases of prolonged or catastrophic illnesses, injuries, and maternity. The amount could be substantial. In maternity, the beneficiary is permitted up to 50 percent of the monthly salary for 100 days. In cases of extended sickness and/or disability, the member is allowed to collect 60 percent of monthly wages for a period of eighteen months. In the self-employed groups, about 74 percent of health insurance plans have "patient cost sharing restoration" programs that pick up part of the mandatory co-payments. These extra financial benefits protect the economic security of every working citizen in the event of a major sickness.

The classifications of the health insurance plans into different groups create discrepancies in age distributions, health and employment risks, and ability to pay. To compensate for this deficiency, government subsidies are provided. Better-funded health programs are required to subsidize health insurance plans at risk financially. In some instances the government subsidy can reach up to 50 percent of benefit payments.

Providers are compensated by the health insurance plans as fee-for-service known as point-fee-system. The point-fee-system is a schedule of fees established nationally and applied uniformly by the Central Social Medical Care Council, consisting and represented by providers, insurance payers, employers, and consumer groups. The fee schedule is adjusted to the type of service. The place where the health service is given, the provider's qualification, and actual cost of the service are of no consequence. As a result primary clinic physicians earn more than specialists. For example, in 1990 clinic physicians earned an average of $22,000 while hospital specialist average compensation was $6,300. The point-fee schedule is renegotiated annually but the Ministry of Health caps the rate of increase.

Extra billing or fees outside the national fee norm is illegal but are common and often ignored. Basic hospital charges, which include board and lodging, medical supplies, hospital staff salaries and equipment depreciations are covered, and nothing else. The inflexibility of the price control provided very little incentive for health provider to do complicated procedures. Doctors seeing more patients and prescribing medications are financially better off than those doing complicated surgical procedures at hospital, such as vascular by-pass surgeries. An average Japanese primary physician sees patients every seven minutes while in America primary care doctors see patients every 20 minutes. This is probably a likely explanation why 30 percent of the total cost of medical care in Japan is spent on drugs. The other is that physicians in clinics can sell the drugs they prescribe, a practice considered to be a conflict of interest in America and completely unacceptable these days.

By population ratio, Japan has more hospital beds than America. The number of hospital beds increased by 25.8 percent from 1970 to 1988. About 90 percent of hospitals with twenty beds or more are classified as general hospitals. They are small, privately owned, and operated as not-for-profit facilities. On the other hand, the average private hospital has 163 beds and a public hospital has 283 beds. The public hospitals comprise 19 percent of the hospitals and accounts for 33 percent of the total bed capacity in the Japanese hospital system.

Japanese view their hospitals as recuperative and not as treatment centers. Traditionally they were used as long-term care facilities. Every large hospital provides not only acute care, but also long-term care. Of the 400 hospitals with more than 500 beds, about 60 percent have only five to seven intensive care beds and 30 percent with neonatal intensive care units. University hospitals are the centers for teaching and training, research, and delivery of tertiary medical care level. There are 131 university hospitals with an average hospital bed capacity of 735 compared with America's 129 university hospitals with 664 beds. Despite the larger numbers of university medical centers and hospital beds, admission rate is 30 percent less than American's and the average hospital length of stay 36.2 days versus 7.9 days in an American university hospital.

Japanese hospitals provide less number of emergency rooms than in the U.S. There are fewer violent crimes, and the free, unimpeded and easy, open access to medical care helped steer away non-emergency cases from the emergency rooms, thus the need for less hospital emergency departments.

Japanese physicians customarily had small-scale practices usually in their homes providing low intensity medical care to their community. More recently, however, they began upgrading their community health services and started acquiring more sophisticated medical equipment to better compete with the hospitals. Some larger clinics set up medical beds in an attempt to keep their patients. When the number of beds is twenty or more they are considered general hospitals.

There is intense competition between private clinics and hospital practices. Physicians of private clinics have no hospital admitting privileges. Once a patient is referred to the hospital, the patient is not referred back to the clinics. Instead she/he is followed in hospital outpatient department. Because of this arrangement, clinic physicians have no incentive hospitalizing their patients and try to put off hospitalization for as long as they can. The rigid price control and other practice barriers explain why there are fewer hospitalizations, expensive diagnostic and therapeutic procedures performed in Japan.

Patients typically are not informed of their diagnoses nor are they given any information about their conditions. Drugs prescribed are not labeled and drug interactions not mentioned. Neither are patients also told if they are part of an experiment. There are no such thing as patient's informed consent and patient's education. Such practices of not fully informing the patient of the diagnosis, treatment risk, etcetera recently were upheld by Japanese court decisions.

With all the shortcomings of the Japanese health system, it is still rated as one of the best in the world. Upcoming problems, however, are large aging populations. With two to three times per capita expenditure for the care of the elderly and the increasing cost of healthcare, Japan is confronted with the question on how to maintain and support its present system.

The Sweden Healthcare System

The *Sweden Healthcare System* is a single-payer socialized medicine similar to the British and Canadian models. It is government run. Every citizen of the country has the right to access a universal healthcare program regardless of her/his ability to pay for it. It is a decentralized government financed healthcare system that is mainly funded through regional and local taxes.

Sweden's healthcare system is one of the top in international rankings. This could be because it is run at a local level where the people are truly involved in decision-making of their healthcare system. It is relatively fast and efficient compared with other EU countries where socialized medicine is practiced.

In the 17th century, families provided healthcare. By the turn of the 18th century, Sweden started small government and church associated care centers, followed by the construction of the first hospitals that were segregated into rural and urban settings. The "provincial doctors" practiced in the rural settings and were responsible for large rural population dependent on the government healthcare programs. Their patients were mostly public (government-subsidized) patients. In the urban settings, the "urban doctors" were not as overwhelmed, and their patients were both public and private. At first these were financed by a combination of state and local funds. In 1733, the national government took over the financial responsibility of the national healthcare and paid both the rural and urban doctors the same. Private healthcare services in all levels of medical care slowly disappeared.

In 1946, Sweden looked into a *National Health Insurance Act*, which was later adopted and initiated in 1955. It called for universal health coverage for all Swedish citizens and was primarily funded by local income tax revenues. Healthcare was controlled and primarily left to the local county and municipal governments. Doctors were allowed to practice privately, although by 1960, 80 percent of physicians worked under government-run hospitals. The *Seven Crown Reform* enacted of 1970 called for doctors to become salaried government employees to operate and manage government hospitals. That ended private healthcare practice in Sweden.

The *Dagmar Reform* of 1984 totally gave control to county councils who decided when and where doctors should work, decreasing the ability of doctors to work extra hours as private practitioners in their free time. Budgets were flexible providing money where it was demanded. There was no incentive to control cost apparently because the politicians wanted to satisfy their citizens.

Healthcare spending rose to 9.3 percent of the Swedish GDP. In an attempt to control healthcare expenses, the Dagmar Reform of 1985 altered the reimbursement formula to one of capitation. To control increasing cost, hospitals and doctors were compensated with the number of patients they served. This led to "global budgets," a system allocating a fixed amount of the fund to each local county. This was the primary attempt to control healthcare expenses. This led to healthcare rationing, which severely affected the quality of care. Healthcare providers were overworked and their work conditions deteriorated, waiting list for medical consultation, diagnostic, and surgical procedures mushroomed; and medical facilities and equipment were limited and outdated and so on.

In 1990, Sweden developed the *"Stockholm Model"* that called for a move from public style healthcare to a market-orientated system. This gave doctors and hospitals more control of their patient care. Quality improved with cost maintained at 8.1 percent of Sweden's GDP. The reform also granted patients freedom of choice over hospitals and doctors. Health facilities, with increase efficiency that provided quality care, attract more patients. Hospitals and physicians were paid on a per patient basis. More patients meant more profit for both hospitals and doctors. In 1992, the *Patient Choice and Guarantee Act* was implemented. It also extends to providers outside the country a guarantee of reimbursement if a Swedish citizen is treated within three months of their diagnosis. Waiting list dropped 20 percent from 1992 to 1993.

The initial success of the reform was short-lived. Waiting lists began to rise in 1994 and so with the cost in 1996 for reasons the Stockholm Model market-oriented system mandates were ignored. The greatest failing of the market-oriented reform was its failure to permit private healthcare providers. With growing concern over the wait lists in 2005, Sweden formulated a new care initiative. It guaranteed, "no patient should have to wait for more than three months once it has been determined what care is needed. If the time limit expires, the patient is offered care elsewhere, which is paid for by his or her own county council, including any travel costs." To this day, the waiting list and overall patient dissatisfaction are ultimate problems of the Swedish healthcare system.

Sweden's healthcare system is organized into three different levels: national, regional, and local. The Ministry of Health and Social Affairs setup the principles at the national level. Guidelines for healthcare political agendas are developed. The minister along with other government bodies, supervise the activities at the regional and local county councils, allocate grants, and periodically evaluate their performances, to ensure the national rules are followed and goals reached.

Sweden is divided into four regions: Gotland, Halland, Skåne, and Western Götaland; and into 290 municipalities and 20 county councils. They are not administratively linked and function independently from each other. The county councils or municipalities are obligated to develop regional projects and the implementation of other national programs. Each county council has a population anywhere from 60,000 to 1,900,000; as a result, they are given the freehand to decide, plan, and deliver the best healthcare needed in the community. Ninety percent of the county councils are also involved in managing other areas of the local community life, i.e., maintaining the immediate environment of citizens such as clean water supply, welfare services, cultural and economic activities, and the building and development of infrastructures.

The responsibilities at the regional level are: to insure healthcare is provided to the local citizens, adequately financed, and decentralized to the local political bodies known as county councils. These political bodies are charged to administer quality medical care to the residents, dental care up to age 20, and to promote good health for the entire local population. They also regulate prices and the level of service offered by private medical practitioners. They are the political representatives elected every four years on the same day of the national general election. The executive or hospital board of a county council exercises authority over management and hospital structure, and make sure healthcare is delivered efficiently. They also regulate prices and level of participation by private practitioners. Private providers must enter into a contract with the county councils otherwise patients are not reimbursed the fee they have to pay their private physicians.

Post hospital care of the disabled and elderly, and long-term care for psychiatric patients recently became the responsibilities of the local municipalities. Other support services are also provided.

They maintain homes and special facilities to accommodate the disabled, elderly, and long-term psychiatric patients. These institutions are probably the equivalent of the nursing homes and hospices in America, and are funded solely by the government.

An estimate of 17 percent of the Sweden's GDP is spent on medical and dental care, higher than most other European countries. The Swedish healthcare system is primarily funded by taxes collected at the local counties or municipals levied by the county councils and municipal governments. In 2003, 72 percent of the county councils' revenue came from taxes, 18 percent as grants from the national government, 3 percent from user-fees, and the remaining 7 percent from other sources.

The cost of healthcare in Sweden was fairly stable till the early 1980s. Since then like any other national healthcare plans, healthcare expenditure has risen at a rate of 2.4 percent per year. Sweden's annual healthcare budget in 2010 was $29 billion.

Primary care for general medicine, emergency treatment, nursing care, and medications are the largest contributors for the increasing cost of healthcare.

The patient's fee schedules as a hospital in-patient are as follow: $22 to see a primary MD; hospital admission fee is $100; hospital stay per day is $12 for the first day then drops to $9 thereafter; consultation is from $52 to $268 depending on county location. Once the patient's maximum annual deduction of around $110 is met, the government pays for hospital care for the rest of the year.

Outpatient fees are somewhat different: $21 to visit a primary care physician with a ceiling of $152 per year, and $52 for a specialist with maximum annual specialist expenditure of $305. Similarly, once the patient's required fees are satisfied the government takes over the payment for the rest of the year.

Prescription medicines are by no means free. Prescribed medications are dispensed only in a network of Swedish pharmacies called the National Corporation of Swedish Pharmacies. It is state-owned and a monopoly, which enables it to maintain a countrywide distribution system. Patients are allowed about $250 yearly deduction. The government pays for any additional drug expense. This network of pharmacies keeps critical medical records especially the patient's diagnosis and medications, so prescriptions can be filled in any pharmacy in Sweden.

Sweden's healthcare is number 23 in the World Health Organization ranking. Although Swede's infant mortality is 2.7/1000 compared to France's 3.4/1000 live births, the long waiting list is perhaps the reason for the lower ranking. In spite, the promises of the Patient Choice and Guarantee Act in 1992, 60 percent of the patient populations wait longer than three months in 2003 for hip and knee replacements, coronary angioplasty/by-pass surgery, prolapsed uterus, hernia, gall bladder, cataract, and etcetera. This long waiting period for treatment is a significant source of dissatisfaction. In 2006, the rate of satisfaction is 74 percent. Today it is 80 percent.

Like many other state sponsored healthcare systems, many challenges confront the Swedish healthcare plan, i.e., issues of rising life expectancy and associated expenditure, waiting period, quality of care, access, and efficiency. To address these concerns, the Swedish government in 2012, reviewed the entire healthcare with the intention of designing a health system that focuses on comprehensive patient-oriented care.

The Swedish people's expectancy continues to rise. In 2010, life expectancy for men is 79.5 years and 83.5 for women. Over 5 percent of the population is aged 80, or over a proportion higher compared to rest of the EU countries. On the other hand, since the 1990s the number of children born has been increasing steadily. By any measure, Sweden is a first world country. Unemployment is 5.6 with a GDP per capita of $41,042. Labor forces work primarily in industry or the service area.

Swedish impression of their healthcare system can be summarized as: 81 percent satisfied with extremely high degree for care of the sick; everyday care is adequate, but the real gap in the healthcare system is in the care of the elders with chronic diseases that lead to at least 90 days waiting. Affluent Swedes, however, find it outrageous that a highly evolved society could abandon 7 percent of its population with no healthcare guarantee at all; the idea of spending 17 percent of GDP seems utterly shocking and irrational.

In conclusion: the Swedish Healthcare System is plagued with many challenges some say it is like a third world country in regards to its accessibility. It is closely operated by the government and ceased to be market orient. Lessons to be learned are: avoid a single-payer government-managed system and let the market determine the true value of healthcare.

The Healthcare System in Switzerland

Switzerland is a country about the size of Vermont and New Hampshire with a population of 7,300,000 people of mixed European descent—French, German, Italian, and Romansch. The country is a federal state consisting of a confederation of 23 cantons (state or provinces) of which three are split to demi-canton (for purposes of convenient representation) acting independently in the governance of the country. Switzerland is a modern society with a very prosperous economy. Swiss GDP in 2004 is $358,000,000,000 translated to about $22,898 per capita; a 15 to 20 percent higher than the big Western European economies.

The Swiss healthcare system is exceptional compared with the other six universal healthcare plans previously discussed. It is perhaps the closest to a free enterprise. As a result it is one of the best in the world delivering excellent healthcare to the Swiss population. It is also one of the most expensive healthcare plans beyond the United States and Germany, spending 11.9 percent of GDP according to Organization for Economic Cooperation and Development (OECD) figures. Like the rest of socialized healthcare plans, the Swiss healthcare system has not escaped the rising cost of providing quality and comprehensive medical care to all Swiss citizens.

Switzerland's healthcare system has its origin in 1890 when the Swiss federal government was mandated to pass a sickness and accident insurance law. In 1911 the federal government legislated the first Swiss Health Insurance Law requiring healthcare insurers to register with the Federal Social Insurance Office and to offer a statutory basic healthcare benefit. Many attempts to reform the health insurance system because of persisting significant financial difficulties failed since its inception. It was in 1994 that a major healthcare reform was finally enacted. It was called the ***Revised Health Insurance Law (RHIL)*** of 1994, and was fully enforced in 1996. The compulsory basic health insurance package was expanded, becoming presently as one of the most comprehensive healthcare benefits offered by any other country.

The Swiss Revised Healthcare Insurance plan was passed to accomplish the following principles and objectives.

- **Patients' preferences:** The population is involved in three-level political referenda that directly influence the decision-making in matters of healthcare choices in the communities. Although the process is seemingly unending the people effectively influence the outcome of any healthcare initiative brought to the public attention for discussion and approval.

- **Decentralization of political power:** The Swiss confederation of twenty cantons and six demi-cantons exercise autonomy in the establishment of healthcare plans in their areas. The cantons and demi-cantons have the authority—within a federal framework—to regulate healthcare activities in their regions from healthcare delivery, accreditation, financing, education, etcetera. The cantonal health minister is charged the responsibility of supervising the healthcare performances in the district and to insure all healthcare needs are adequately met.

- **High Degree of Competition:** Incorporated in the RHIL is a high degree of competition in the Swiss healthcare industry. The Swiss are free to choose from the variety of registered healthcare insurance companies which best serve their needs and whether to buy healthcare supplemental insurance or not. Patients are likewise allowed to choice their own doctors and have the freedom to change their doctors as they feel appropriate to meet their healthcare requirements. As a matter of practice to increase business, insurers and healthcare providers compete for patients as clienteles. Healthcare insurers may also selectively establish networks of healthcare providers for their own customers.

The freedom of the Swiss public to buy and choose their own individual health insurance and their healthcare providers/physicians, and the health insurance companies' ability to compete and align with managed-care institutions or form physicians networks to attract more clients, convinced Harvard Business Professors Nancy R. McPherson and Regina Herzlinger that the Swiss healthcare is a consumer-driven healthcare system and ought to be emulated. Both professors are of the opinion that Switzerland's consumer-driven healthcare plan has accomplished universal healthcare coverage and constrained cost with significantly lower price than the United States healthcare system without the appearance of a government-controlled healthcare program.

- **A Unique Mix of Private and Public Financing:** The Swiss Healthcare Insurance system combines both public and private funds in a somewhat unusual way contrasted to other social healthcare programs do. As a result the expense from public funds is the lowest in the European Union. Financing of the healthcare from tax revenues, direct payments from compulsory, and supplemental healthcare insurances have decreased in the last twenty years, although the actual healthcare expenses have been steadily increasing.

  When analyzed as to who actually assume the cost the Swiss healthcare system, private households bear 68 percent, followed by cantons contributing 15 percent, the federal government at 7 percent, businesses put in 7 percent and 3 percent for local authorities.

The Swiss Revised Health Insurance has three components. They are the compulsory basic (a very comprehensive health insurance coverage) better known as social insurance, voluntary supplemental health insurance, and the sickness, old age, and disability insurance.

RHIL guaranteed enrollment of every Swiss citizen in any of the social insurance of choice regardless of age and/or health risk. It obligated either the federal or cantonal governments to review and establish that the healthcare services offered by the insurers registered under the social insurance category meet exactly what the law dictates. Insurance premiums in this group are federally negotiated and regulated but not permanently fixed. The premiums among the different insurance companies in this classification vary depending on the amount of co-payments required and/or the *franchise-ordinaire (FO*/minimum deductible) or *franchise-a-option (FAO*/higher deductibles) are mediated and determined annually at the cantonal government levels and are uniformly applied through out the canton. In actuality, the premium is community rated. It is applied equally without discrimination for every citizen in the canton. Insurance premium rate is based on age at the time of enrollment, and not of risk—a reason why many Swiss stay with the same insurance company for most of their lives.

The RHIL opened the possibility of innovative funding for the ever-rising cost of healthcare. With managed-care introduced in the 1990 it quickly grew with the passage of the healthcare law of 1994, offering lower premiums in lieu for restricted choices. So far two types of managed-care have evolved—***HMO-type*** policy and ***General Practitioner Network (GPN)*** system.

The HMOs are the leading social insurance. They are mostly owned by health insurance companies and are operated as staff-model facilities with salaried physicians and other healthcare providers. With a standard FO policy, they are usually 10 percent to 20 percent lower than the other health insurances. The GPNs are located mainly in smaller cities. They are a network of primary physicians agreeing to act as gatekeepers for the health social insurance companies. Physician's compensation in the GPNs is by capitation. Physician providers in the network are required to share 50 percent of either profits or losses of the plan. Physician's losses, however, are capped annually at ten thousand Swiss Francs (SwF 10,000). The GPN premium for a standard FO policy is roughly 5 percent to 15 percent lower than the regular social insurances.

While the Swiss government regulates the health insurances, RHIL also compelled all Swiss citizens to purchase their own individual social insurance health policy. Those who are less well off are given premium reduction and subsidies paid directly to the insured if her/his premium is 10 percent or more of income. Those who refuse to buy their own basic health insurance are severely penalized. This mandatory social (healthcare) insurance is to satisfy the Swiss healthcare philosophy of "collective responsibility."

RHIL also allowed the Swiss population to voluntarily purchase supplemental healthcare insurance. Approximately 40 percent of the population subscribe to this voluntary supplemental health insurance which usually covers healthcare services not covered by the basic social insurance, e.g., patient's freedom of choice of any hospital and hospital accommodation, dental services not related to medical illnesses, treatment by chief physician, etcetera.

The third component of the Swiss healthcare system is the mandatory old age, sickness, and work-related/accident disability healthcare insurance. The health benefit package is similar to the social insurance coverage. The difference between the two is the manner of funding. The sickness-old-age-disability group of insurance funds is financed by compulsory employer and employee contributions based on the employee's income, consisting of about 7 percent of the total Swiss healthcare program.

In 1998 there were twenty 23,679 physician and dental practitioners in Switzerland. About 13,357 or 56 percent were independent office-based private doctor/dentist practices—of these 36 percent are in general practices and 46 percent are involved in specialty practices. Ambulatory patient cares are treated similarly as in the United States in hospital's outpatient clinics, polyclinics, self-style-self-financed HMOs' medical/multi-specialty centers, and independent private doctor's medical offices.

Health/medical/dental cares are paid on a fee-for-service manner. Services are itemized immediately after each doctor's office visit and an invoice is presented to the patient. The patient's total payable bill is adjusted according to the patient's deductible, co-payment and type of health insurance policy. Physician/healthcare provider's fees are nationally determined annually by a point-value scheme by associations of healthcare insurance funds and professional organizations, and are setup in a fee schedule, which is then implemented nationally. In 13 cantons, healthcare providers are allowed additional compensation by "freedom of prescription" which enables practitioners to prescribe medications and are allowed to dispense the drugs they prescribe in their offices. Thirty-three percent of provider's income in these cantons comes from such drug prescription revenues.

Managed care is becoming popular in Switzerland as an effective cost containment measure. There are about 3,792 doctors belonging to the GPN system covering 350,000 people and ten Swiss staff-model HMOs with 140 physicians insuring 98,400 individuals. HMOs are multi-specialty clinics with general practitioners and specialist in internal medicine, gynecologist, nurses, physiotherapist, etcetera. Patients requiring other specialty care are referred to other specialists and hospitals to either the doctor's recommendation or the patient's preference. With restricted choices permitted by managed care, there is presently some question and disagreement as to whether clients are satisfied with the HMO and GPN healthcare plans or not.

The hospital infrastructure of Switzerland is generous—with 406 hospitals in 1997. Five are considered university hospitals for medical education and postgraduate training for physicians in the different specialties of medicine, surgery, laboratory, and medical research. There are 5.6 available beds per one thousand of the population. Swiss hospital length of stay is comparatively higher than with other EU countries. This perhaps explains why the country's hospital expense is one of the highest, making it one of the most expensive healthcare systems in the world. Unlike in other European countries, this excessive hospital bed capacity and length of hospital stays to date has not been addressed adequately.

Both publicly and privately owned cantons or local health authorities regulate hospitals. The number of hospital beds, hospital accreditation, in-hospital care, hospital funding, reimbursement of hospital providers and hospital payments by health insurances, and any other healthcare program is decided and publicly subsidized at the local government levels. Hospitals and healthcare providers outside the accredited list are not compensated for medical services rendered and are financially devastated. Privately-owned hospitals included in the canton's accredited list are eligible for social insurance reimbursement. New private hospitals may encounter great difficulties getting in the list of accredited hospitals in the area. Health insurance companies favor in-hospital to outpatient care because in-patient care is partly paid by the local public.

The federal government has no authority in such matters of healthcare. The tight control of the hospital system and hospital care by the local cantonal health authorities provides no incentives for innovations that will optimally and economically reform the Swiss healthcare system. Swiss healthcare and economic experts including economic Professor Zweifel contended that although there appears to be some competition in the Swiss healthcare system, there is really none.

The cantons and local authorities finance approximately 50 percent of the hospital capital investments and operating expenses of the public and not-for-profit hospitals. The other 50 percent comes from a daily flat fee paid by the social insurance companies for in-hospital medical services. This daily rate is annually negotiated and determined on a canton-wide region between the associations of social insurance companies and organizations of healthcare providers. Hospital healthcare providers are salaried with possible additional pay coming from treating patients with supplemental healthcare insurances.

There are three types of hospital accommodation required in all public and private cantonal hospitals in Switzerland. These are the ward (allgemein), semi-private (halb-privat) and private (privat). Like in the post WW II era to the 1990s in the United States of America, the wards are big rooms with more than four beds. Rooms with two beds are semi-private room and those rooms with a single bed are classified as private hospital room. The basic social insurance entitles the patient for an in-hospital admission to a hospital ward in their canton of residency. Depending on the patient's supplemental health insurance benefit, the patient has the option to select her/his hospital and/or type of hospital accommodation.

In every healthcare plan, deficiencies are easily identified. In the Swiss healthcare system one big problem is the drug benefits. Only one-third of prescription drugs are covered by the social insurances and to a certain extent—depending on the benefit package purchased—by the supplemental insurances after the patient pays a 10 percent of cost co-payment. In most instances the patient pays the full price of the medications. Perhaps this is the reason why there is a great push by the federal government advising the public to request physicians to prescribe generics and for the pharmacists to substitute for generic drugs.

Akin to the other social health systems, the Swiss is experiencing similar problem of serious escalation of healthcare expenditures. Even though public satisfaction is high, many consider it very expensive and arguably not very cost effective. Because of rising healthcare costs, policy makers and critics of the system are asking questions as to what level of healthcare do the Swiss want, how much are they ready to pay, and/or will they accept healthcare rationing.

Which healthcare system do you think will best fit America? Should America adopt a healthcare plan similar to any of the countries' healthcare systems reviewed? Canada, Sweden and the United Kingdom health systems are primarily financed and dictated by the government with the citizens having no financial responsibilities. Long waiting lists in every level of healthcare, however, are rampant.

France, Germany, and Japan on the other hand impose obligatory employer and employee salary contributions of 14.2 percent average percentage for the three to finance the workers' health insurance premiums tightly controlled by the government. Delays in diagnosis and treatment are not as bad as in Canada, Sweden and U.K. but the health systems are highly regulated. Switzerland in contrast mandates every citizen to buy his or her own insurance with severe penalties if it is not done. Insurance premiums and healthcare funding are controlled at the various levels of the local and federal government that changes are almost impossible to come by.

Again which system would you prefer? Will your choice work effectively here in the U.S.?

VI

Discussion and Comments

Evolution of America's Healthcare Crisis

On the evolution of the healthcare crisis, it seems that the gradual intrusion of the government post-World War II into the healthcare system, and which culminated in the mid 1960, fueled the uncontrolled rise in healthcare cost in the United States. Noticeable increases were seen two years after the introduction of the Medicare and Medicaid programs in 1965.

In an attempt to slow down the growth of healthcare cost, the federal government financially supported and endorsed the development of managed care in the early 1970s, particularly HMOs. This led to the introduction of a healthcare concept never before tried in the U.S. healthcare system. This permanently altered the practice of healthcare in this country. Managed care rapidly flourished. HMOs just like their government counterparts imposed their own monetary regulations, which proved quite advantageous for them. Healthcare costs seemed effectively controlled initially.

It drastically changed the business dynamics of the healthcare system in favor of the managed-care companies. Profits began to be made. However, instead of using their profits to decrease healthcare cost through lower insurance premiums, insurance premiums instead kept increasing, and at the same time doctor's reimbursements fees were adjusted downward. Managed-care companies generated huge profits that were used to finance investments not directly related to patient care. The business was so profitable some saw the opportunity of converting not-for-profit managed-care companies into for-profit corporations.

The ***Physicians Health Plan***, a Minnesota not-for-profit physician managed HMO was converted to a for-profit health plan. It is presently the second biggest health insurers in the country: ***United Healthcare.*** Smaller managed-care companies were slowly swallowed by bigger HMOs or other healthcare insurance companies through mergers that continued to the present, resulting in the steady elimination of competitors and the formation of very big, politically active, powerful and influential managed-care institutions.

With the great financial successes of the healthcare institutions from the clinics/hospital to the health insurers, board of directors of healthcare companies found it necessary to increase CEO compensations to keep them "on board." According to IRS documents in 2003, the CEOs of the top six tax-exempt hospitals received an average of $1,200,000 plus other generous benefits of up to $5,000,000 a year. A survey by the Chronicle of Philanthropy showed that executives of not-for-profit hospitals were the fifth highest paid CEOs in the world of not-for-profit organizations. The salaries of the CEOs of the largest for-profit hospitals, ranged from $2,000,000 to $16,000,000 annually. The annual median compensation of hospital administrators with more than $1,000,000,000 yearly revenue was at $786,000; $486,000 for hospitals with less than $1,000,000,000 revenue; and $302,000 for hospitals with less than $200,000,000 revenue. These level of wages prompted K. B. Forbes, the head of an advocacy group for hospital reforms, to state, "We can understand someone making several hundred thousand, but when they're pulling in seven figures, they're acting like they are the NBA of the hospital sector."

Looking at the yearly income of the health maintenance organizations (HMO) and other managed-care organizations (MCO) executives, the average compensation of the biggest twenty HMOs and MCOs in 2003 was $15,000,000 according to Ron Pollack, Families U.S. president. United Healthcare CEO William W. McGuire held an unexercised stock option worth more than $530,000,000 in 2002. The highest paid HMO executive in 2004 was Norman C. Payton of Oxford Health Plan earning

$76,010,825, followed by the chairman of Well Point, Leonard D. Schaffer at $21,765,532.

In two most recent news articles in the *Star Tribune*, United Health Group CEO had a total compensation of $125,000,000 in 2004 and an unexercised stock option worth $1,600,000,000 in 2005. Other United Health senior management officers have exercisable stock options—COO for $663,000,000; general counsel at $60,000,000; SVP at $50,000,000. Last but not least are the ten outside United Health directors who cashed in their stock options totaling $159,000,000—an average $ 15,900,000 per director.

All top healthcare insurer CEOs are paid salaries that are extreme and the median of their excessive compensation is exactly in the middle. "Let them make multimillions in business, fashion, entertainment—but not by depleting valuable healthcare dollars," alleged by Dr. McGarvey, a retired Indiana otolaryngologist.

A professor of economy at the Duke University, Frank Sloan, PhD made the following observations: "If someone is earning $10,000,000 a year or $20,000,000 a year that seems quite excessive, given what we are experiencing with the healthcare. It just creates an upward spiral. I'm not really ready for regulations, but I think we need to worry about these multiples (of CEO pay vs. average wages). These multiples are discouraging to people. A lot of us work because we love the work, we just do it out of love, and we don't have to be paid $7,000,000 to do it."

Alarmed by the abuses of HMO/MCO, Ralph Nadar on July 15, 1999 wrote Senate Majority Leader Trent Lott, Senate Minority Leader Tom Daschle, Speaker of the House Dennis Hastert, and Minority Speaker of the House Richard Gephardt, and recommended:

- HMOs and MCOs must be held legally accountable when necessary medical treatments are delayed or denied.

- Apply the Employee Retirement Income Security Act of 1974 (ERISA) with no exceptions to all,.

- Only doctors make the decision and determine what is medically necessary to their patients and not by corporate bureaucrats.

- Conduct evaluation of HMO/MCO procedures and policies by independent third party review groups.

- Cap the salaries of HMO/MCO CEOs that service Medicare, Medicaid or otherwise receive tax money.

- Authorize and encourage healthcare consumers to band together to form Consumer Health Action Groups across the nation.

One good sign noted in the statistical section is the ability of smaller HMOs and MCOs to offer comparable healthcare coverage at much lesser premiums in contrast to bigger and well established health insurance companies (HICs), HMOs and MCOs; perhaps because big companies have much higher overhead costs and administrative expenditures. **This observation is one of the important reasons why the two-step bidding process is necessary to control and decrease healthcare expenditures.**

It is important to underscore the trend in the statistical data presented. It truly illustrates how much our healthcare had grown into disarray. The many contributing factors are by no means simple but rather multifaceted issues that unfortunately are also interrelated to each other. Solutions therefore are hard to come by, and would need the introspective inputs by healthcare experts, government policy makers, academicians, professionals in other fields, consumer groups, and other interested groups or individuals who have different perspectives that can be considered and incorporated in finding answers—thus the need of involving intense public interest and debates on issues confronting a universal healthcare plan. Good examples are the various problems so far identified and discussed.

The ramifications of our mounting healthcare expenditures are far and wide, seriously affecting our well being as a nation and as a people. The effect on our economy is considerable. It has undermined our ability to compete at home and in a global economy. An automobile manufactured in the U.S. has a $1,400 cost disadvantage compared to cars built abroad because of the added employee's healthcare benefits. The same assessment was expressed a few years ago—by G.M. Canada's CEO Michael

Grimaldi, co-signed by Canadian Autoworkers Union President Buzz Hargraye—in a letter sent to the Crown Commission considering reforms of Canada's 35-year-old national health program to avoid accelerated healthcare costs in Canada. "The public healthcare system significantly reduces total labor cost for automobile manufacturing firm, compared to the cost of equivalent private insurance services purchased by U.S.-based automakers." It is "vitally important that the publicly funded healthcare system be preserved and reviewed, on the existing principles of universality, accessibility, portability, comprehensiveness and public administration" and institute "updated range of services." CEOs of Canadian Ford and Daimler Chrysler also wrote similar endorsements of the Canadian Medicare to the Crown Commission. The fact is, American healthcare is far more expensive—about 66 percent more than the Canadian Medicare.

No wonder American automobile manufacturers are having a hard time selling their products not only in America but also in other countries. GM and Ford have been losing billions of dollars the last few years that the only way they can stay in business is to close plants and lay off tens of thousands of auto workers.

The truth is the high cost of our healthcare has affected every segment of the economy adversely. American manufacturers are having a tough time selling their products here and abroad. Closure of manufacturing plants in various industries and lay offs are a frequent occurrence. That means not only loss of income but also of healthcare, retirement benefits, and possibly employee's homes and their other assets, adding farther to the number of people without health insurance and retirement plans. One can only imagine what a person or family goes through after several months of unemployment.

Then there is the outsourcing of jobs and the flight of American companies to the third world countries to avoid the high cost of doing business here in the United States. There are obviously many factors affecting high labor cost here in the U.S. One thing for sure is the high cost of employee healthcare benefits. Companies all over the United States are having difficulties with their healthcare benefit programs that about 50 percent of both small and medium-sized companies can no longer provide healthcare coverage to their employees. Among large companies, there is increasing awareness of this problem as they find much difficulty meeting expenses for their employees' healthcare benefit programs.

A more direct effect of high healthcare cost is the number of U.S. bankruptcies reported in 2000. About 50 percent of bankruptcies filed were from hardworking middle-class Americans who had at one time before and during their illness been covered by health insurance. Their healthcare bills were so high they could not afford to pay the amount they're responsible; ending up in bankruptcy.

Similarly, countries like Canada, England, other European countries, and Japan that adopted national healthcare plans controlled healthcare cost successfully in the beginning are now experiencing same difficulties the U.S. healthcare is having. This is particularly true in all the seven industrialized nations reviewed with national health insurance plans giving universal access to healthcare. There is a plausible explanation why healthcare costs in these countries are starting to explode. Foremost is the total control the governments had on the healthcare programs they have.

Too much government control effectively abolished the forces of checks and balances inherent in a free market economy. It effectively eliminated free market forces. Competition disappeared. The single-payer government exercising much regulation rendered the system incompetent and wasteful. Fraud and abuses increased. Wasteful practices unchecked and persisting. Patient healthcare services reduced. And because the government was always urged to do more by public demands, the government is tempted and enhances its medical expenditures and benefits for its citizens.

The government's tight control of the healthcare systems in Canada, United Kingdom, France, Germany, Japan, Sweden, and to a certain extent Switzerland clearly illustrated a form of indirect healthcare rationing. In Canada for example the availability of services that are clinically important are limited. There is overcrowding, long waiting lists, and unnecessary delays in healthcare facilities forcing many Canadians to seek healthcare outside the system. Patients are unable to see their primary physicians for several weeks because of overbooking in clinics and hospitals. It is common for patients to wait six to twelve months for elective surgeries such as coronary by-pass surgeries or hip replacement because of long waiting list.

A CT scan, MRI, or MRA and/or other medical procedures that are expensive and requiring state-of-the-art apparatus are not readily available. Instead these are delayed for several months because of the limited number of available equipment. This means usually going south to the U.S. or using private health insurances to have medical care in private clinics and hospitals, which incidentally are illegal but ignored and tolerated. The same situation is going on in England. It is, however, worse. In the Japanese healthcare system, it appears there is availability of enough hospital beds and modern medical equipment. But because of the way the Japanese government had structured and restricted the healthcare system to correct inefficiencies, hospital beds are used more for long-term care rather than for acute care. The type of involved medical care commonly practiced in the United States is regulated and discouraged.

Rewarding primary care physicians who can quickly see patients every seven minutes and prescribe more medications indirectly does this. This practice of relaying more and more on drugs in managing patient's diseases explains why the Japanese healthcare system spends 30 percent of all healthcare expenditures on medications. Well trained specialists in contrast, who do complicated and risky diagnostic and therapeutic procedures in Japanese hospitals, are compensated about 72 percent less than the primary care doctors. This one-sided payment in favor of the primary physicians resulted in much less diagnostic and high risk, intense and complicated medical treatments performed, i.e., coronary angiograms, vascular by-pass surgeries, cholecystectomy, prostatectomy, hysterectomy, exploratory laparoscopy, eye lens implantation, radiation therapy, pace-maker/defibrillator implantation, colonoscopy, CT scan, MRI, ultrasounds, etcetera. On the average, Japan does approximately 33 percent of what U.S. healthcare providers do in diagnosis and treatment for elective but life threatening medical conditions.

French, German, and Swiss healthcare systems are a fair compromise between the Canadian, British, Japanese, and Swedish systems, and that of the United States. It is in these healthcare systems that the government's heavily controlled national health insurance schemes allow and accept the use of private health insurances, as primary and/or supplemental health insurances are well coordinated. Canada, England, Japan, and Sweden appear to be moving toward allowing private insurers as an essential part of their healthcare. It looks like there is growing desire on the part of these healthcare systems to move away from the inflexible management style the government has on the healthcare system. In fact, Canada's Supreme Court recently ruled that it is unconstitutional to forbid private health insurance when the public healthcare system can not deliver reasonable service needed by the people. This Supreme Court ruling possibly opens the wider use and acceptance of private healthcare insurance as a complementary part of the Canadian Medicare.

On the other hand, the U.S. healthcare is moving rapidly towards what the seven other countries are trying to remedy and/or avoid. The ObamaCare, in particular, accelerated the process of transition from a relatively free market directed healthcare to an intensely government controlled one; as more provisions in HR3200 are added.

Despite claims of universality in the healthcare systems of Canada, England, French, Germany, Japan, Sweden, and Switzerland, it is worth mentioning again that a significant segment of their population are excluded in the national health system and are provided healthcare by a different health plan, or completely ignored.

The Organization for Economic Cooperation and Development (OECD) considered France as the "the best health system in the world" and is ranked first in overall performance. England is in the eighth and Canada in the 30th place in overall performance. Japan is ranked in the 10th place in fairness, first in overall attainment of goal and 10th in overall performance. Germany is the sixth in overall fairness, 14th in goal attainment and 14th in overall performance. The U.S. is in the 54th place for overall fairness, 15th in goal attainment, and 37th in overall performance. U.S. definitely lags way behind these countries, particularly France.

The overall performance ranking by World Health Organization is based on the following: good health; responsiveness of the country's healthcare plan to the population's expectation; and fairness in funding the healthcare system.

- France: 1
- Japan: 10

- England: 18
- Switzerland: 20
- Sweden: 23
- Germany: 25
- Canada: 30
- United States: 37

Although these countries are ranked way ahead of the U.S., significant reforms by all seven governments are underway to address their service deficiencies and expanding healthcare expenditures.

Which of the different countries' healthcare systems presented would you like to have for the United States of America—France or maybe Japan or England? Would you prefer saving at the expense of convenience, timely diagnosis, and treatment, knowing well that these countries are beginning to have difficulties financing their healthcare plans in the very near future?

Paraphrasing the conclusions of David G. Green and Benedict Irvine in their research on *Healthcare in France and Germany: Lessons for UK* published by the *Institute for the Study of Civil Society 2001*:

- The government should not take the role of a single-payer for it will lead to rationing;

- The government should not impose upon in the health system a single-payer for it will trap consumers in situations of bad services, and decrease incentives in raising the standard of care in the healthcare system;

- The government should not link persons with employers because it will be difficult for the consumer transferring to a better healthcare facility;

- *Government policies should recognize the unique nature of healthcare; that it is in most part a cultural, ethical, moral, and societal obligation for the government to provide, and to a lesser extent, an*

ordinary consumer good that eventually everybody needs;

- Self-sufficient members of the society must be totally cost conscious and the dependent members be equally price conscious as dictated by circumstances;

- People dependent on the government for their medical care should not receive inferior medical services;

- Healthcare systems should not and could not differentiate the social and economic status of the participating member/citizen.

Government regulations and laws are meant to give the structural organization and the rules of interaction, so there is orderliness, fairness, equity, and equality among the participants. It is not supposed to preclude any legitimate and reasonable activity. It is worth remembering that when deregulation occurs—especially in areas tightly controlled by inflexible regulations implemented both by the private institutions and government sectors—the dynamics of a free market economy is at its best, and the consumers always win.

Free market economy in itself is not without fault. Such flaws if anticipated or identified can be readily corrected and improved upon. The recognition of existing and potential problems that the healthcare system may encounter in the future is therefore vital, in preserving and protecting the integrity of any healthcare system that may be established soon or later on. Some of the problems of the present healthcare system identified had been discussed, but need further elaboration in some areas. The topic on *"abuses by patients"* presents specific difficulties to practitioners and requires good documentation of the patient's medical visit with her/his doctor.

A standard method for documenting a patient visit to the doctor in the U.S. is SOAP, as previously noted. It is a method of documenting a patient's visit to her/his doctor consisting of the patient's subjective symptoms, objective findings on examination, assessment or diagnosis made, and the plan of managing the patient at the time of the visit. Most physicians (if not all) discuss their plan of care with their patient; a very different approach compared with the Japanese. Questions and/or feedbacks from the patient are encouraged during the encounter. In majority of instances the patient agrees with the doctor. Some, however, are too unreasonable. No amount of logic can dissuade them from whatever they think should be done. To avoid further confrontation and waste of time, poor patient's satisfaction surveys, reprimand from superiors and supervisors, possible future malpractice litigation, etcetera, some physicians accede to patients' demands. Educational seminars and training sessions on "*How to Deal with Difficult Patients*" at times may work. The time-honored business phrase of "the customer is always right" is an oxymoron that has no place in the practice of medicine, and must be removed completely in administrative policies of the government and corporations. The use of generally accepted clinical guidelines by healthcare providers give some protection but not sufficient enough in some cases.

Rules that will allow doctors to practice freely their profession and to protect the healthcare provider from abusive patients must be a part of a healthcare initiative. Patients insisting on the newest and expensive drugs outside a pre-approved drug formulary should be required to pay higher co-pay or the full retail price of the drug. Specialized testing and procedures, unless clearly indicated in the opinion of a primary care physician or ordered by a specialist, should also cost the patient higher co-pay or better yet the entire bill of the procedure or testing. Written rules protecting healthcare providers from patient abuses will go a long way in curtailing unnecessary healthcare expenses labeled as "defensive medicine."

The professional healthcare fee for services here in the United States is 29.1 percent of the total healthcare expenditure. The primary dilemma in the fee schedule is the lack of a standardized basis for pricing should be made. As a result there is extensive variation in fees around the country. The same thing is true with what healthcare consumers pay for drugs, non-durable and durable medical supplies, dental, and other local or institutional healthcare services. The difference may be as much as three to six times more than the lowest price offered in any other community.

A recent survey of generic drug cost among pharmacies in the Twin Cities revealed that some charge from 300 percent to 600 percent more compared to the wholesale price for the drug.

Pharmacist's professionally fee for filling/refilling prescriptions is 7 percent to 8 percent of the total drug cost. The industry's usual reasoning for charging so much is because of overhead expenses and the stocking of many drug inventories in the pharmacy. It sounds rational at its face value. But then, why is it that grocery stores still make reasonable profits in spite of keeping large quantities of perishable goods? How about the retail stores, like Target, Wal-Mart, or Best Buy selling products that become seasonally out of fashion or easily rendered obsolete still make good profits with 150 percent to 200 percent markup on the products? The most likely explanation is abuse of the system in the absence of real competition in the healthcare industry.

This great variability in the cost of medical services, medications, and other medical products is one of the major reasons why we have healthcare predicaments today. Pricing and charges have no rational basis. Weaknesses and lack of proper guidelines in various areas of the present healthcare system provide unprecedented opportunities for abuse, exploitation, fraud, and waste. Although very distasteful for many healthcare practitioners, some form of standardization in pricing is a welcome necessity if we're to make sense of the ballooning healthcare expenditures. A logical method is to adopt the **Resource Based Relative Value Units** system or **Relative Value Units (RUV) for short**. It is presently based on practice expenses and professional liability insurance multiplied by a conversion factor determined by the Centers of Medicare and Medicaid Services. RVU should be a standardized method of compensation or pricing that attempts to combine and quantify the effects of the cost of practice and malpractice insurance liability premium, the risk and complexity involved in decision-making, the time spent, and the actual work itself. For medical products including drugs, the RVU should be calculated from the cost of R&D, cost of production, and marketing only to healthcare professionals, product liability insurance premium, and length of patent and residual revenue after the patent.

However, RVU system is still inadequate in its latest form and application. It requires further development and improvement under the supervision of Healthcare Security Agency and the Department of Health and Human Services, and perhaps the

Department of Commerce and the National Institute of Health to widen its scope to include every possible type of healthcare service, procedure and medical products. This RVU method of reimbursement should be uniformly applied throughout the nation—without any exceptions—to prevent widespread variations in personal healthcare expenditure.

While the Japanese healthcare system allows primary providers much better compensation than their specialist colleagues who perform complicated procedures, the opposite is true in the United States. Here risky and complicated procedures are a valued part of medical care that frequently saves many lives. They are done by skilled doctors using modern instruments, and are usually time-consuming, thus very expensive to do. Specialists doing procedures here in the United States, unlike their counterparts in Japan, are compensated very well so that there comes a time when the economic incentive of doing more may come into play. As a result, specialists are much better paid than their primary care colleagues. There appears to be a significant inequality in both systems where the primary care physician is favored in one system and the specialist is at an advantage on the other system. A fair compromise between these two systems is a concept that could be achieved by the RVU system.

Since there are many variables included in the calculation of an RVU, the system should be closely monitored and the value of the RVU periodically/annually adjusted and determined—up or down—by a commission under the auspices of the Healthcare Security Agency, represented by members from the consumer, healthcare profession, business community, and representative of the nation's state Healthcare Fund agencies.

The most expensive medical care is that given to the elderly. It is controversial to cap and/or limit the healthcare benefits of the elderly who live beyond their "life expectancy." It is, however, also a very logical solution if we are to slow the rising cost of healthcare. The elderly should be managed conservatively with proven, effective methods to promote maximum comfort and independence. Aggressive medical and surgical treatments should be avoided at all cause, e.g., cardiovascular by-pass, transplants, skeletal prosthetic surgeries, and the like in medical management, etcetera. The needs of our elderly or the dying must be approached with compassion and provided in a conservative manner that accommodates the immediate needs and respects the end of life with charity, dignity, and honor.

Doctors must stop "playing God." Instead they should meet with the patient, her/his immediate relatives, and/or care givers to discuss *advance medical directives*, and negotiate an acceptable plan that is sensible and believed the most appropriate medical management for the patient and those immediately involved in her/his care.

On the other hand, people who desire or expect to live healthily with good quality of life well beyond their projected life expectancy should have the option to provide themselves with added protection that covers aggressive treatments and/or other forms of healthcare therapy that may be outside the usually considered conservative medical managements. It is the responsibility of the elderly person and/or the family to decide whether to buy supplemental coverage or not—just like what is done presently with *Medicare Supplemental Insurance Coverage* and/or *Extended Healthcare Coverage.* Another option or possibility is for a state to elect providing funds for such coverage from the unspent SHF or SHRF, keeping in mind that perhaps the reason why other nations have longer life expectancies than the U.S. is that they do less and are not as aggressive in medical managements as their American counterparts.

Again it is worth mentioning that more treatment does not necessarily indicate better outcomes. The seven national health services evaluated herein provide fewer healthcare services than the United States of America. Still they are considered some of the best healthcare systems with lower infant mortality rates and higher life expectancy, definitely better than the U.S.

Another group of patients to be considered for healthcare benefits capping are the non-compliant patients, who, after repeated admonition from their healthcare providers to follow treatment protocols recommended for their care, continue to practice unhealthy lifestyles without regard or consideration to the consequences of their actions. Yet, they often expect the best and most aggressive cure available to be done, including repeated cardiovascular by-pass surgeries, transplants, implants, and other expensive medical and surgical care presently available.

The solution to this particular problem is to have an independent *Medical Ethics Committee*—preferably at each major community healthcare institution in the county—consisting of a medical ethicist, four healthcare providers from different specialties; two from the consumer group and two from the business community; one social worker and one person representing religion.

Individual members of the Medical Ethics Committee shall not be subjected to any legal liability or action for decisions made by the committee—in good faith.

The function of the ethics committee is to make decisions as to the appropriateness, and/or approve or disapprove medical treatments wanted/requested/demanded by non-compliant, chronically, and debilitated patients, based on patient's and/or family's wishes, patient's quality of life, lifestyle, ability to comply with recommended medical managements, availability of social programs, and the socioeconomic and financial costs to the community where the patient lives. Any controversial medical management issue that comes out in the course of a patient treatment should be discussed and resolved by the local Medical Ethics Committee.

Another fault of an unchecked free market is uncontrolled profit motivation. Businesses in the healthcare industry presently can ask increases in their prices practically as far as the market will bear, as the saying goes. Take for example the healthcare insurers. A yearly occurrence is the increase in premium two to three times greater than the rate of inflation. These annual healthcare premium increases are easily justified through accounting schemes in favor for the companies. Enormous profits are made, abuses take place, CEOs and management teams paid fat salaries, and non-direct patient care investments disguised as administrative or capital expenditures.

Imagine this repeated several times at the various levels of the healthcare service industry, year after year. Can the healthcare companies be blamed playing this game? Of course not. It is an inherent goal of business to make as much profit as possible. How much depends on "what the market will bear." The industry pretty much carried out whatever business plans they had, realizing that everybody needed healthcare services. Healthcare users were captive consumers. And so it was in the last few decades of steady increases in healthcare costs in America.

Everybody involved in the healthcare businesses in the United States must realize that healthy, productive Americans are an enormously valuable resource of the country. Taking business advantage in a system that everybody needs is pure greed, inconsiderate, and should be unacceptable and condemned. Let us all agree that keeping Americans healthy or treating their diseases with the best medicine offered inexpensively is an ethical and moral imperative each business in the healthcare has to accept, and not as a commercial opportunity to take advantage and exploited for big profit.

The persisting perception among business leaders that the healthcare is the industry where unlimited profits can be made regardless of consequences ought to stop. This kind of attitude must change to avoid the economic disruption high healthcare expenditure is causing the U.S. economy. Our healthcare resources are finite. It would be to everybody's advantage, especially for businesses to realize that healthcare is not the place where unlimited profit should be made. It is the business where equitable but not outrageous profit is possible.

Powerful tools of the free market have—not present in the current healthcare—to check the excesses and unwarranted profit motivation of companies are to install sizeable healthcare funds and re-introduce the concept of bidding in the healthcare business. These large health funds constitute tremendous and compelling business leverage for both the buyers and sellers of health services, and a force to be reckoned with by the sellers to seriously consider their prices in the process of bidding. It becomes an "all win or all loss" situation for the sellers (bidders). Winning a bid means the business of a very large population with the possibility of an equitable profit. Losing the bid on the other hand signifies loss of business for a long time which most companies can ill afford.

The re-introduction of free market forces in the healthcare system through competition, business leverage and negotiation; formation of networks of exchanges, alliances and/or other healthcare players/businesses should successfully control, most if not all, the problematical factors ailing the present U.S. healthcare system.

New studies show that education and prevention extend realistic quality of life even among those who have advance diseases. Redirection of healthcare resources to programs proven to prolong longevity and promote good health must be an important focus, consideration and goal of a healthcare system.

As noted, one of the problems is a nonchalant attitude we have about our health, especially the young. Changing attitude is very hard to do. It will take a lot of time, money, and effort. Preventive measures must start at a very young age to inculcate the importance of good health. As suggested, the health science studies should be a mandatory part of our school curriculum starting from the first to the twelfth grades. It should be a required subject just like reading, writing, mathematics, and science. Subject matters to be studied should include the importance of personal hygiene, nutrition, exercise, rest and relaxation, avoidance of stress, mental health, common infectious diseases, serious communicable and sexually transmitted diseases, sex education, pregnancy, alcohol/tobacco/drug use/abuse, driver's education, first aid and resuscitation, alternative medicine, and introductory courses in anatomy, physiology, biochemistry, and other health and medically related subject matters deemed important. The subject matter should be taught at the appropriate age or grade level of the students. Courses have to be designed and determined by qualified experts in the fields of healthcare and education.

Funding for the health science studies program may be wholly and/or partially from both federal and states with the monies coming from the Healthcare Security Fund and State Healthcare Fund allocated for education. Likely identifiable outcomes of the suggested health science studies are a health-conscious population, healthier lifestyle, decrease and prevention of more serious diseases, and an easier transition for students into higher education planning careers in the healthcare profession.

The OECD grading is unquestionably a strong argument by healthcare advocates why the United States of America should adopt a health system similar to the seven countries reviewed.

Which health system best fits the U.S., considering the multicultural background of our society? The rankings most likely were based on data collected as to fairness, goal attainment, and performance of each healthcare system, without any considerations to other factors or events affecting the outcomes of the research. To mention just two of the numerous factors in the U.S. possibly affecting the results of the OECD ranking are wars, illegal immigration and our lifestyle.

Illegal immigration in the U.S. is up to one million individuals a year. Estimates put the number of illegal immigrants in the U.S. at 11,000,000 to 22,000,000 or more, a large number by any standard. Illegal immigrants are from third world countries that are poor, seeking either economic opportunities or medical care here. Once they are here they hide and are undetected, belonging to the underground economy. Contrary to the common arguments given by pro-illegal immigrants, illegal immigrants working in this country are paid agreed-upon hourly rates in cash. Federal and state taxes, social security, and Medicare deductions are not made. Their employers, whether they are big corporations or individuals, do not keep records for their illegal immigrant's employment. They are unrecognized and "invisible" until they become seriously sick and look for medical care. Their poorer health and more serious medical condition due to lack or delayed medical care will certainly affect our statistical results.

The other important factor is our lifestyle, i.e., our work ethics, overeating, inactivity, and obesity, use of tobacco, alcohol and illegal drugs, and driving habits. Let us examine our lifestyle closely as a country.

We as a people are envied all over the globe. We are the most productive society in the world, and because of that, we are the most prosperous nation with a very high standard of living. We donate billions of dollars to other countries in foreign aid and other humanitarian help. We deploy our troops all over to protect countries and maintain global security and stability, costing us lives and hundreds of billions of dollars in expenses. In spite all our goodwill efforts, we are the most criticized, disliked, hated, and misunderstood people.

We are the most overworked, tired, burnout, stressed-out, and sleep-deprived people on earth. We work the longest hours—about

40 percent more than the French, Scandinavians, and the English people—and take the least/shortest vacation compared to our counterparts in other developed nations. We are hard working sometimes to the detriment of our health. Our dedication to our work and work ethic are frequently misunderstood by others as a love of money, material things, and greed.

To support families, two incomes from the husband and wife are necessary. At times the husband and/or wife have one and a half to two jobs to keep the family financially solvent. That means our children are unsupervised after school that lead to gangs, alcohol, and/or drug abuse, crimes, etcetera. To adults this may develop in lifestyle abuses, e.g., alcohol and drug use, illicit sex, overeating, family violence, etcetera to relief the stress and distress of daily living. Unfortunately, our political and liberal thinking leaders fail to recognize or acknowledge what the majority of us in the middle class are going through. To top it all up, some liberal politicians and activists, including a former U.S. president, feel ashamed and guilty America is not doing more for the world. We should help people from underdeveloped nations who are impoverished and devastated by natural disasters, social and political upheavals, but we should take care of our people first and not give away all the assets we have—to the needy countries of the world.

Blame us for the incompetence, corruption, cruelty, fanaticism, and greed, power-hungry, etcetera of other nations' leaders, government, and repressive cultures. It is always America's fault for the failures of foreign governments and problems around the world. We've been branded as imperialist. And as the only superpower, we accept the moral obligation to help and lead but not to impose our will or govern others, and we are hated for it.

Yet we can not provide some of the very basic necessities of life to millions of our citizens, such as adequate food and shelter, healthcare, education, and employment. Where is our priority? Is taking care of the world our primary agenda as a nation? Or is taking care of our own citizens our first priority to remain spiritually, morally, ethically, economically, socially, and politically strong? Again with our stressful lifestyle and its ramifications our health statistics are negatively impacted. Our government's domestic and foreign policies have to change if our healthcare cost is to be effectively decreased and controlled.

Then there is the issue of malpractice litigation in the U.S. In the societies examined, malpractice is not as popular as here. If malpractice is filed the rewards are usually a small fraction of what is awarded in the U.S. Perhaps the reason is cultural differences. People from other countries have high regard of their physicians. Health provider's medical decisions are respected and seldom, if at all, questioned. Their argument is that doctors study medicine for several years and then train for many more. Therefore they are much more knowledgeable than the patient and know what they are doing. The opposite is true in the U.S. Some, however, attribute the rarity of malpractice suits in Canada, U.K., France, Germany, Japan, Sweden, and Switzerland to the universal access of healthcare in these countries and regulations they have regarding malpractice suits.

According to the *National Fee Analyzer,* 30 percent of medical bills contain erroneous entries in charges and/or coding. A medical billing auditor Nora Johnson said, *"More than 90 percent of the bills I review are either wrong or padded beyond belief"* and *"Would you believe a thousand dollar toothbrush?"* The errors are often in the billing departments of large medical clinics and hospitals. These mistakes in medical billings are not isolated cases but are happening all the time across the U.S.

An effective way to counter these abuses is for the patient or healthcare consumer to be mindful of her/his medical bills by requesting itemized bills, keeping notes of what procedures and/or services were done and the doctors' visits. A careful review of all medical bills should be undertaken especially when being admitted through the emergency room where double billing frequently occurs, and any discrepancies immediately reported to both the healthcare provider and the healthcare insurer. Periodic unannounced independent audits by *State Attorney General/SHF agency* should likewise be conducted and violators severely punished with fines and incarceration.

The elimination of paperwork no doubt will greatly help the business and practice of medicine. This would mean the standardization of billing, insurance, and other necessary business forms into a format acceptable by all concern. Healthcare practice/business will be greatly enhanced with the incorporation of the electronic medical record and a national healthcare identity card. Both business operations and healthcare records could be easily managed by electronic transfers, which would immensely simplify access of healthcare anywhere in the U.S. or the world.

However, strict protection of the patient privacy rights should be a top priority. This may call for the revision or modernizing the *U.S. Penal Code* to include severe punishments in the use without the proper authorization of the person, or commercialization of private personal information. Subdividing one's personal information into different dossiers, i.e., medical dossier, sociopolitical dossier, economic dossier, should help. Subcategories of the sociopolitical dossier should include general information dossier, and law enforcement dossier. These dossiers should be handled by separate private and/or government agencies and are outlawed to be cross-referenced.

Creation of an Ideal Universal HealthCare Plan for America

The proposed universal healthcare plan for the United States addresses above-mentioned deficiencies. If passed and enforced as a federal healthcare plan, these concerns will be corrected by the forces of checks and balances of a free market economy; without any doubt.

There are many good ideas already published on how to fix our present healthcare dilemma. Most of the suggestions, however, are fragmented solutions. They do not deal with the whole spectrum of problems confronting the healthcare system. Majority are more like patchwork attempting to correct one aspect of the healthcare at a time. So far nothing has truly changed the healthcare system and the expanding healthcare cost continues on.

A good and popular example backed and advocated by many experts in the health and finance areas is the *Health Savings Accounts (HSA)*. President George Bush signed HSA into law on December 8, 2003 in its present form. It is a combination of a high deductible comprehensive health insurance and a savings account similar to the *Individual Retirement Account (IRA)*. The health insurance has a minimum deductible of $1,050 and a maximum of $5,000 out-of-pocket healthcare expenses for an individual, and a $2,100 minimum deductible and a $10,000 maximum out-of-pocket expense for a family. Annual contribution of $2,600 for an individual and $5,150 for a family are tax deductible. HSA can accumulate yearly if not spent like an IRA. Withdrawal of the HSA fund for medical care expenses is not taxable. It acts as a supplemental for medical expenses.

The concept of HSA is rapidly becoming acceptable by both employers and employees. In a survey by Melton Bank in 2005 360 employers with an average of 9,000 employees, about 66 percent of employers contribute to employees' HSAs. Estimated growth in the number of HSAs is 1,600 percent increase from 2005 to 2008, and the corresponding projected dollar value is from $282,000,000 in 2005 to $6,000,000,000 by 2008.

The rationale behind the HSA program is good, but again it does deal with one specific problem in our present healthcare. It gives individuals the opportunity to manage her/his/family medical expenses and be aware of the actual cost of healthcare. Admittedly, this should help diminish the rising expenditure of healthcare in the U.S. One disadvantage predicted is the tendency of people not to seek medical care on a timely basis and/or for preventive care. The other problem is not everybody can afford to make the contribution, and it is applicable to only a small segment of the population. Medicare recipients are excluded. There is also the possibility that an individual or family in some cases can easily deplete the maximum allowable out of pocket medical expenses year after year. This is especially true among patients who develop later chronic illnesses with recurrent exacerbations and/or persons who had been seriously injured in automobile and/or work accidents after they're enrolled with an HSA program, where the automobile and/or workman's compensation insurance coverage had run out.

A new experimental version of the HSA for Medicaid recipients is the ***Health Opportunity Accounts (HOA)***. The reasoning is the same as the HSA. It will expose Medicaid beneficiaries to the real cost of their medical care, and motivate them make decisions for their healthcare. The elderly and disabled are excluded. Participating states can contribute as much as they wish to the HOAs and the federal government matches up to $2,500 for adults and $1,000 for children annually. Like the HSA, the unspent HOA fund can accumulate just like an Individual Retirement Account. The same concerns also apply to HOA.

Similar HSA or HOA healthcare government program had been implemented in Singapore. While the Singapore healthcare system appears awfully successful, one must not forget that Singapore is a tiny country of 4,350,000 relatively very young population, compared with that of the United States of 310,000,000 much older population.

With the increasing healthcare financial pressure, governors are increasingly concerned, and the labor unions supporting state legislation requiring large corporations to provide healthcare benefits. Lately, a number of state governors expressed interest in introducing bills in their legislative bodies to reform their healthcare systems. This interest in fundamentally changing their healthcare varies in degrees and approaches, and which appears to be gaining momentum. Labor unions on the other hand are pushing for legislations in 30 states requiring large employers to make payroll contributions to their employees' healthcare plans; features of the French, German, and Japanese national healthcare plans.

Another is Medicare D. It was great in intention but too complicated in implementation and expensive at $40,000,000,000 yearly projected budget. Worst of all, Medicare D was denied the ability to bid for better drug prices with the different pharmaceutical companies—a clear evidence of the effects of influence peddling by lobbyists, and the graft and corruption permeating the different branches of the federal government. With the power to negotiate successfully eliminated, expect drug prices to rise in the coming years. Eventually the $40,000,000,000 yearly budget will not be sufficient and there'll be clamors for increases. Medicare D is projected to grow to $100,000,000,000 yearly in five to ten years. In the meantime the pharmaceutical industry will be bloating with profits. Are the American people allowing this to continue? One thing good that is emerging is the intense competition among insurances trying to enroll as many as they can, and has actually helped more than 80 percent of seniors with their drug cost.

The silent majority must speak out and push for drastic restructuring of the present healthcare system that provides universal access for every American, but is market and consumer-driven.

The various developments in the healthcare reform debate are too complicated and confusing. The solutions offered by the various reform debates and healthcare initiatives are too fragmented without unanimity and/or uniformity. It can easily lead to different forms and/or tiers of healthcare—one for the poor, another for the middle class, a third tier for the rich, famous and influential, a fourth tier for the federal employees, a fifth tier for the armed forces, and a sixth tier for the executive, legislative and judicial branches of the federal government—to exaggerate and to put it bluntly: that may just be worse than what it is today.

The establishment of a universal healthcare plan for America was based on the general principles discussed. Perhaps the most significant is the consolidation of resources earmarked for healthcare. This will enable the federal government to fund each state's healthcare requirements. It is the best mechanism by which U.S. can provide healthcare coverage to each citizen with no added cost. Since the distribution is according to a per capita formula, the amount of money received by each state is enormous by any measure; giving the state powerful leverage to bargain for better prices with the healthcare business community. With the rules spelled out, the roles of the federal government become limited. The government is prevented from interfering with the business aspects of implementing healthcare delivery; allowing free market dynamics to determine the true commercial values of healthcare services. The same can be said with the state governments.

One significant obstacle that comes to mind, however, is the unwillingness of our elected federal officials to give up their healthcare entitlement, which is far better than the general public's available health coverage. If this is what stops this proposed plan from becoming, then the public can be generous and give concession to our government officials letting them retain the federal employees health benefits; similarly financed and administered as it is today. The armed forces and other federally funded healthcare programs with the exception Medicare and

Medicaid can be treated in the same manner. Even with the exclusion of the Federal Employees Health Benefits, and the likes from the proposed universal healthcare plan, the pooling of monies from other sources already identified will still generate vast sums of money for the Healthcare Security Fund sufficient to finance the healthcare needs of the rest of the country.

The other possible obstacle is the refusal of the members of the U.S. Congress to distribute the Healthcare Security Fund equally among all the citizens of the United States of America. The total money disbursed to each state should be based on an equal per capita formula and not by any other means—as used in the distribution of social security and Medicare benefits—where Californians and Floridians receive higher benefits than citizens in other states in the country. To adopt such a policy is shortsighted and will ruin the general principles from which this healthcare initiative is based upon. It is also the highest form of citizenship discrimination.

The Canadian Medicare, British National Health Services and Sweden Healthcare System are funded primarily by general taxation. In other countries, the government obligates the employer and employee payroll contributions to finance the national health insurance systems, and to a small extend by private funds. In the French National Health Insurance System employees contribute up to 20 percent of payroll, Germany's Healthcare Insurance System, the contribution of employees is from 12 percent to 13 percent, and the Japanese Health Insurance Act assesses employees up to 9.5 percent of payroll. The Swiss people buy their own healthcare insurance coverage. Supplemental healthcare insurance coverage is bought by 10 percent of Canadians and English, 11 percent of Germans; 12 percent of the French population and 40 percent of the Swiss. Additional healthcare expenses in these countries are co-payments from 10 percent to 30 percent of healthcare services; with Japanese paying the most.

In the U.S. the senior's healthcare is funded by compulsory payroll deduction of 2.9 percent of employee's gross income for Medicare; the employer contributing the same percentage of payroll. The very poor persons receive their healthcare from the Welfare/Medicaid federal and state programs funded by general taxation, and the 47,000,000 working poor or not so poor Americans who can not afford to buy health insurance are without coverage. Voluntary employer and employee payroll contributions cover the rest of the population. And those in employment transition continue their healthcare coverage by paying the entire premium that is usually quite high.

The question is: "Will the American people tolerate and/or accept a universal healthcare system financed singly by compulsory employer and employee payroll taxes as proposed?" The employer's contribution had been suggested and discussed. How much are employees and self-employed individuals willing to contribute for such a universal healthcare system? If paying a little bit more would mean an uninterrupted and a worry-free healthcare coverage for everybody, employers and employees would probably not have any considerable opposition to the concept if employers' contributions and employees' payroll deductions are within reason.

After all, most of us are already paying substantial amounts in out of pocket expenses for our healthcare. For the employers this would be ideal and should be embraced by them. Companies will have predictable healthcare expenses, which are tax deductible, and free them from periodically negotiating healthcare coverage with their employees' unions—a rather difficult, messy, and expensive undertaking. The employees and self-employed on the other hand are assured reasonably priced, comprehensive, and continuous healthcare coverage no matter what, the freedom to choose her/his healthcare providers, and timely medical care.

French, Germany, and Japan have an average employee payroll deduction of 14 percent; not including the cost of their supplemental private health insurance premiums. In Switzerland, since the people buy their own insurance coverage and the health insurance premium varies widely up to 50 percent differences, it is rather difficult to actually determine what each person pays. Some insurance premium average estimates roughly put it at $2,790 per person annually. This estimate translates to about 12 percent of the average per capita income, and does not include cost of supplemental health insurance carried by 40 percent of the population. Is a *Healthcare Security Fund Tax* of 10.5 percent to 12.5 percent of a person's gross income too much to ask?

Americans are a generous people. They understand the importance of controlling the skyrocketing healthcare cost, especially if the proposed concept means an access to healthcare by every American.

To simplify collection, Congress may consider combining Medicare tax and Healthcare Security Fund tax of 12 percent to 14 percent of total gross income as one healthcare tax. Of course these numbers are arbitrary and have to be determined by actuarial calculations and have to be shown as sufficient to fund a universal healthcare plan as introduced. These (author's) estimates are likely to be on the high side. *A more realistic percent contribution for all individual employers and employees, and self-employed persons with income is maybe 8 percent; the figure cited as penalty for non-electing employers in ObamaCare, is reasonable Healthcare Security Tax to replace the Medicare portion of FICA. It seems reasonable to believe the tax revenue generated will be sufficient to finance also Medicaid.*

The portions of the HSF and SHF allocated for health and medical education, research and development at the federal and state levels are meant as supplement to the already budgeted funds for these purposes. These funds should greatly help our country devote more resources in the fields of health education, research and development of new treatment and technology.

Provisions of the intended universal healthcare plan of importance are the ***Americans Living Abroad Healthcare Agency*** and the coverage of Americans traveling aboard. Americans traveling abroad have to buy additional health coverage to protect themselves from unforeseen medical needs while in foreign countries. Medicare does not cover retirees living outside the U.S. presently. They purchase their own health insurance in the country where they live or go uninsured, paying healthcare expenses they incur as they go. Unless the medical care is catastrophic, most of these Americans can afford to pay for their healthcare because of the much lower healthcare cost in other countries, and the high value of the dollar especially in developing countries. Not covering Americans living abroad and/or traveling in foreign countries is not fair and a big injustice to these American citizens, especially to the Medicare recipients who have contributed to the system during their productive years. They are also taxpaying Americans. It will be a gesture of discrimination and treatment of these Americans as second-class citizens if they are excluded.

Illustrated Implementation

Implementation of the universal healthcare plan is relatively simple. Once the federal and state governments have collected and disbursed the funds, the governments' roles are fundamentally fulfilled. Unlike the national health plans of other countries reviewed, where the healthcare systems are heavily controlled by so many rigid government regulations, this universal healthcare proposal practically prevents interference from the government. Free market dynamics take over the healthcare delivery system—in all its possible levels of operation averting unnecessary expenses, delays, and other inefficiencies that may arise.

The two-step bidding process is an essential part of the proposed healthcare system to promote and encourage competition. The lowest secretly submitted bid publicly announced and a one-week (or so) delay before the open public reversed auction is conducted, will give the other bidders a chance to reconsider their bids, and again re-enter the competition for the county's healthcare. Understandably, the competition to reduce the price will be intense, considering the loss of the county's business will mean a sizable business loss for the losing health insurers. The same argument can be made to all the various business operations in the healthcare system. The formation of consortiums/exchanges, alliances, and networking among the different businesses engaged in the healthcare undoubtedly will result in increased leverage in bidding/negotiation for lower prices of healthcare services and products.

The two-steps bidding provision is expected to be strongly opposed by the healthcare business community. It is encouraged that if ever this health initiative is ever take-up in the U.S. Congress, that the American people really have to be involved advocating for this healthcare initiative.

Nelson A. Paguyo, MD

Advantages of the U.S.UHP

Embodied in the United States of America Universal Healthcare Plan are the specific roles of the federal and state governments and the private sector. The governments' roles are limited in the collection and distribution of monies for the healthcare system, and the formulation of general guidelines that will allow free market forces to work effectively as it should be.

The U.S.UHP allows and encourages—if not forces—the different players of the healthcare system to freely compete, be successful and efficient that will prevent excessive increases in healthcare cost; and determine the fair value of healthcare services and products under constantly changing business and market environment.

It frees American employers from periodically negotiating healthcare benefits with their employees and their unions. With their healthcare cost expressed as a fixed percentage of employee's total gross income, it is a predictable business expense and becomes tax deductible; they'll be in a better position to predict cost and compete in the global economy just like the other industrialized countries of the world.

Everybody pays and forces all participants to realize that U.S.UHP is their own, that they have a significant stake in the system and to act responsibly keeping down the cost of healthcare. It expediently eliminates illegal aliens and non-resident or non-refugee aliens in using and taking advantage of our healthcare system.

U.S.UHP will provide uninterrupted and equally comprehensive healthcare benefit/coverage for all Americans— young or old, employed or unemployed, Medicare or Medicaid recipients, healthy or sick, home owners or homeless, and rich or poor—at no additional cost and with the possibility of reasonable savings. It offers all Americans, especially our most vulnerable citizens, a worry-free and easy accessibility to America's healthcare. Last but not least, it practically grants financing for research and development, health education, and other healthcare initiatives both at the federal and state government levels.

Disadvantages of the U.S.UHP

Healthcare Security Fund is based upon on the percentage of the individuals' earnings and employers' contributions—including Medicare. As such the amount collected will fluctuate and be subjected by the economic conditions of the nation and possibly of the world. This is the reason for both the HSF-RF and SHRF funds are designed to last ten years to insure adequate funding in the event of economic depression in the United States and/or around the world. At any rate with the competitive market principles imbedded in the plan, it is likely the national healthcare expenditure per year will not exceed 16 percent of the U.S. GDP.

It may tempt the federal and state governments to install too much regulation that causes the healthcare system to become rigid or inflexible, becoming just like the other countries with national healthcare insurance systems.

Life Expectancy

Life expectancy shall be defined as the statistically determined average of the number of years a person is expected to live based on death certificates in the United States as it is now generally accepted. In the past few decades, life expectancy had risen to its present level for man at 76 years and woman at 78 years. Persons living beyond life expectancy mean their internal organs are beginning to fail and/or are already failing due to old age. There is no known method of treatment at the present that can reverse the process of aging. Eventually, the organs of the body fail, and death occurs.

There is, however, two medically recognized approaches to take care of the diseases of the elderly beyond life expectancy. One is the conservative medical treatment approach supplemented with appropriate use of dietary adjustments and supplements, exercises, medications, physical therapy and rehabilitation, integrative medicine, comfort care, and management of the dying, easily achieved by Healthcare Teams. The alternative treatment is the aggressive medical and surgical

and easy to implement. It is relatively antifraud and abuse proof. It can be design so your face, name, address, and account number are the only information contained in the healthcare card. Swiping or scanning the bar code in your healthcare card at point-of-service will allow the computer to access your data base file (showing only your name, address, card number, picture, and fingerprint) verifying that the NHID card presented is tampered free and the bearer is you and you are the person you claim you are. The information contained in the NHID card is just enough to expedite healthcare transactions but not to other sensitive data—such as one's personal information, medical record, etcetera.

In countries like Canada, England, France, Germany, Japan, Sweden, Switzerland, and other nations where national healthcare are successfully carried out, the National Healthcare ID card—called in Europe as the *European Health Insurance Card (EHIC)*—is well accepted and a routine part of the healthcare scene. In fact, with open trade policies among the European Countries, health insurance companies are permitted to cross national borders to insure any European citizen a healthcare smart card had been developed in 2004. The countries of Europe immediately endorsed the idea, and began extending healthcare reciprocity to all European citizens who are traveling. The smart card is now widely used and accepted in the European Union as an essential part in the practice of medicine. There seems to be no reported misuse and/or abuse in the use of the smart card in Europe.

In the United States the Center for Medicare and Medicaid Services urges physicians to enroll and get a *National Provider Identifier (NPI).* It is aimed at eliminating other forms of physicians' identifiers. It is a form of a national healthcare provider ID, which is mandated by the government in order for a medical provider to get reimbursed properly and timely for services rendered to Medicare and Medicaid recipients. By requiring physicians to have NPI, and promoting EMR and other electronic standards, the federal government hopes to reduce the administrative cost on healthcare providers and convince them to give up paperwork/claims for good. Such a system undoubtedly will simplify and make life easier for providers, but the proper use of physicians' information should be secured and protected against fraud. If the federal government is endorsing, encouraging and ordering healthcare practitioners to have a National Provider Identifier, why then can't the general public have a National Patient Identifier to facilitate the business and practice of medicine.

Without a National Healthcare ID, any form of a Universal or National Healthcare Plan can not be properly enforced. It should be considered just like any license to drive an automobile, practice one's trade or profession, or use of a credit card. The personal data required in the U.S.HPA application form are common entries, and frequently asked personal information by credit card companies and other financial institutions for commercial loan applications— with the exception of the picture and fingerprint of the registrant, which is seriously being considered by financial institutions as part of the application in the not so distant future.

Consider the present, with personal dossiers being collected by different institutions (mostly commercial), citizens really have no effective means of control on how their personal information is collected and used. The result is extensive and unregulated collection and use or misuse of personal data, by anyone, that most often far exceeds the original intended purpose of collecting personal data.

Personal privacy can be effectively protected if legislations with severe penalties are made to outlaw the use of personal information without the knowledge and written consent of the person. A good start is the ***Health Insurance Portability and Accountability Act of 1996 (HIPAA).***

In addition, a person's dossier must be divided into three main categories, e.g., medical, economic or financial, and sociopolitical. A general information dossier and law enforcement dossier maybe subsections of the sociopolitical category. These data categories must be handled by three different and separate institutions, and not allowed to be cross referenced and/or cross reference with each other, to safeguard sensitive personal information. Only the individual person can authorize the use of her/his data in the different categories as needed in medical care or financing for example.

Nelson A. Paguyo, MD

Funding of the Proposed National Healthcare Plan

A significant portion of the healthcare expenses comes from Medicare in the tune of $527 billion. Add to that is $260 billion for Medicaid. These two government programs represent a total of $787 billion or 31 percent of the total healthcare expenditures in 2010. Both programs are financially compromised. Medicare is expected to go bankrupt by 2016 according to the Trustees of Medicare because of the ObamaCare. Medicaid, on the other hand, will never go bankrupt. The federal government finances ninety percent of it, and 10 percent is by the state. Both Medicare and Medicaid will be reduced as entitlement expenses are adjusted down with the present fiscal debt of $16.7 trillion.

FICA tax rate is 13.3 percent of payroll. The Social Security portion is 10.4 percent. The Medicare part is 2.9 percent. With the ObamaCare, the Medicare portion will go up to 3.8 percent. The two should be separated and Medicare payroll tax changed (as suggested) to the Healthcare Security Fund Tax (HSFT).

In 2011, according to Kaiser Family Foundation, the estimated revenue for Medicare was $486 billion—42 percent from general tax revenue, 37 percent payroll tax, 13 percent beneficiary premium, 3 percent of Social Security tax, 1 percent from state and 4 percent from interest and others.

The total personal income of all Americans in 2010 and 2011, reported by the U.S. Bureau of Commerce (Bureau of Economic Analysis) were $12.31 trillion and $12.95 trillion respectively. If $12.95 trillion is taxed as the HSFT at 10 percent, it will generate $1.295 trillion of revenue—enough to cover both Medicare and Medicaid expenses. It eliminates also the tax burdens of seniors from the 13 percent Medicare beneficiary premium, and 3 percent Social Security tax.

The employee should not shoulder the 10 percent HSFT. Rather it is divided as 5 percent payroll tax for the employer and 5 percent for the employee.

In other countries with national healthcare plans, the employer and employee each contribute to the healthcare coverage of the employee. In France, employers and employees are assessed each 20 percent of payroll; Germany 12.5 percent; Switzerland 12 percent; Sweden 10 percent; and Japan 9.5 percent. In addition to the payroll tax employers and employees pay in the seven countries reviewed, approximately 30 percent of their populations also carry supplemental private health insurances; calculated about an average of 12 percent of their income.

Beginning in 2013 the Medicare portion of FICA goes up to 3.8 percent—a difference of 1.2 percent from the 5 percent recommended.

The other possibility is for both employers and employees to pay each 10 percent of payroll. This will generate $2.59 trillion, enough to cover the entire national healthcare expenses. It will effectively eliminate other healthcare benefit responsibilities employers have to their employees, and to the employees a significant reduction and/or elimination, and consolidation of all other healthcare expenditures. Contentious periodic negotiations between management and unions (for health benefits) will disappear.

Is it too much to ask the employers and employees to pay 10 percent tax on payroll that will guarantee all Americans a healthcare plan that is high quality and comprehensive, reliable and uninterrupted, easy to implement and user-friendly with mechanisms for funding health education, research and development, and health reserve funds, continuous and uninterrupted, promotes economic stability and America's global competitiveness, and lastly, a healthcare plan that does not exceed 16 percent of GDP?

In summary, the proposed U.S. Universal Healthcare Plan can be funded by any of the following: 1) the pooling of all monies earmarked for healthcare expenditure from the various federal and state government health programs and private sources; 2) a FICA Medicare part increase to 5 percent payroll tax by both the employer and employee, plus all the monies from the private sector designated for healthcare expenses; 3) a FICA Medicare part of 10 percent payroll tax by both employer and employee.

VII

Concluding Statements

Medical science in the United States is the most advanced in the world. Healthcare, however, is costly and beyond the reach of about 47,000,000 people of the population. It is highly priced and continues to escalate to a point average Americans will not be able to afford in the next five years or so. Problems (13 in all) responsible for the excessive spending in the healthcare system had been identified, discussed, and solutions to make medical care accessible and cost-effective to every American had been presented. Unfortunately, the newly passed HR3200 law did not provide solutions to the 13 main problems driving the growth of healthcare expenditures. In fact, ObamaCare made it worse.

An overview of the healthcare systems of Canada, England, France, Germany, Japan, Sweden, and Switzerland showed different management styles. These are not, however, acceptable with the American way of life. Nevertheless, these schemes brought varying degrees of successes. The flaws of the various health systems were recognized and considered, and possible solutions included in the proposed universal healthcare for America.

The rising cost of healthcare has adversely affected Americans. Businesses are increasingly having troubles competing with foreign products. Manufacturing plants are either closed or moved to third world countries to escape the high cost of labor, magnified by soaring healthcare expenses. Fifty percent of bankruptcy filings by average Americans are due to medical expenses patients can not pay.

Abuses at every level of the healthcare delivery system exist. Worse of all, the business of medicine are perceived as an industry where unlimited profits can be made. Many are taking advantage of the situation. The entire United States population is considered basically a captured market with significant competition fundamentally non-existent

ObamaCare, when fully implemented will ruin our healthcare system. In 2013, with the ObamaCare fully in place, America will be spending $2.683 trillion (ObamaCare $163.40 billion plus $2.520 trillion for the present national healthcare expenditure). A conservative 3,000 percent increase from 2013 to 2022, brings the total national healthcare expenses to $80.49 trillion, more than 5 times that of our $15.094 trillion GDP in 2012. If the healthcare expenditures grow like Medicare and Medicaid, by 2022 the total is $186.92 trillion. These projections, based on available government figures are frightening. How can America afford such an expensive healthcare? To avoid this "healthcare cliff" the present healthcare system must be drastically changed.

The proposed healthcare reform calls for certain key concepts the author has known as significant elements of a Universal Healthcare Plan in order for it to work. These are: 1) the collection and pooling of all financial resources designated for healthcare expenditure into a massive single fund by the federal government from the different federal and state agencies, employers and the individual taxpayers; 2) the distribution of the federal healthcare fund to the different states based on state (per capita) population to make sure that all Americans have funds for their healthcare; 3) the allocation of the healthcare fund (from the federal government) by the state to provide each of its citizens healthcare coverage; 4) the administration and management by the state of the healthcare fund, either by self-administering the fund and/or buying commercially available healthcare coverage from insurance carriers or managed-care organizations, done by a fair two-step process of competitive bidding; 5) the promotion and formation of alliances, consortiums, and network of exchanges among the various healthcare industry players to lower cost in a free market environment, with the least interference from the government; 6) the identification of serious problems that ail the present healthcare system and the recommended solutions appropriate to these problems; 7) the strict enforcement of federal mandates in the healthcare plan; 8) the issuance of the national healthcare ID card; 9) the creation of funds for health education, medical/healthcare research and development; 10) the establishment of healthcare reserve funds; lastly, 11) the acceptance of responsibility and ownership by the American people of the Universal Healthcare Plan.

The coming together of all these elements should establish a viable and maintainable healthcare plan that covers all the citizens of the United States of American; unlikely to exceed 16 percent of the GDP. The government and private sector are designed to cooperate in this health initiative. It is meant to aggressively, comprehensively, and radically restructure the current healthcare system. It is financially feasible and reasonably competitive that gives every American a continuous access to their healthcare system free of worries and concerns.

The principles presented here maybe applicable—with slight/some degree of modification—to other countries desiring to have a sustainable universal healthcare plan.

It is the hope of the author, that this Universal Healthcare Plan herein brought to your attention, can be used as a framework by the U.S. Congress and other countries to develop and legislate much needed reforms in the healthcare system of the United States of America and other countries. The author encourages the reader to help advocate for such badly needed healthcare reforms. They are requested to call their elected officials in Washington, D.C.

PLEASE CALL YOUR STATE U.S. SENATORS AND REPRESENTATIVES AND DISCUSS THIS WITH THEM—because the present U.S. healthcare plan and the ObamaCare included will surely bankrupt this country unless existing problems of the present healthcare system are properly corrected. Send them a copy if they do not have it. If enough copies are circulated and read, it should spur a serious national debate that will lead for the enactment of a truly UNIVERSAL HEALTHCARE PLAN FOR AMERICA that is affordable and sustainable.

VIII

Strategy for Advocacy

Help advocate to dismantle and then rebuilt from start the present U.S. healthcare scheme. The construction of the new healthcare system must be based on free market economic principles, especially that of competitive bidding to make it affordable and sustainable for generations to come.

Please campaign for healthcare changes—urged and endorsed in the book that all Americans deserve—to as many relatives, friends, acquaintances, professional colleagues, co-workers you know, particular your healthcare providers. We need a critical number of Americans to help advocate for healthcare reforms; who will contact and encourage their elected officials in Washington, DC to legislate healthcare laws similar to what the book recommends.

If you are outside the United States of America and a citizen of another country, you can ask your government to enact healthcare laws based on the suggestions in the book. Your healthcare costs are also becoming expensive and unaffordable. The rapid increases in healthcare expending of your country is forcing your government to limit, healthcare benefits/services that lead to major problems encountered in the seven countries reviewed.

Let's not delegate our individual responsibility to the next person, or to the next generation. Do whatever you can and what you think is right for you—to have this going. Do not underestimate your effectiveness, personal conviction and contribution. It will make the difference in this worldwide healthcare advocacy.

Spread the message for healthcare reforms—here in the US and in your country.

REFERENCES

- Nelson A. Paguyo, MD. A Framework for a Universal Health-Care Plan (Solution to a Health-Care Crisis). © Copyright 2004

- Nelson A. Paguyo, MD A Universal System of Identifying. © Copyright 2005

- Nelson A. Paguyo, MD Healthcare for all Americans: Healthcare Crisis USA—A Comprehensive Solution. © Copyright 2007

- Various Articles in the Internet I Found Researching Pertinent Topics—Many were cited in Healthcare for All Americans.

- Television and Radio News 2004 to 2012.

- Newspaper, Magazine and Circular Articles 2004 to 2012

  Medical Conferences attended 2000 to 2006.

  Michael J. Carson. Healthcare for All Americans: Healthcare Crisis USA – A Comprehensive Solution. The Midwest Bookreview, Reviewer's Bookwatch: January 2008.

ABOUT THE AUTHOR

Nelson A. Paguyo, MD is a graduate of the University of Philippines, College of Medicine; class of 1963. That same year, after passing the Educational Council for Foreign Medical Graduates qualifying exam, and Philippine Medical Board exam, he migrated to the U.S. to pursue his postgraduate training. He became a citizen of the United States in 1977.

Dr. Paguyo completed his residency in internal medicine and fellowship in cardiopulmonary at the University of Minnesota (UM); then became the lead investigator to a NIH research project—Chronic Bronchitis Emphysema Clinic at the UM Mount Sinai Hospital Division in Minneapolis. To help support his growing family while in training, he "moonlighted" as a house and emergency room/department (ER/ED) physician at Divine Redeemer Memorial Hospital (DRMH) in South St. Paul, Minnesota.

After successfully passing the Minnesota Medical Board in 1972, Dr. Paguyo opened a solo practice clinic in internal medicine in So. St. Paul, Minnesota. He continued working as an ER doctor to help finance his new practice. At the request of the hospital and the medical staff, he began setting up the Respiratory Therapy Pulmonary Lab Department at DRMH. He was appointed chief to run and manage it. That same year, he organized a group of licensed physician to provide coverage at the hospital emergency room. The group was incorporated later as the Southern Metropolitan Physicians Professional Association (SMPPA). He became its CEO.

SMPPA was the first of its kind in St. Paul that availed licensed doctors to cover emergency departments.

The concept was so new in the early 1970; very few hospitals around the United States had it. As the idea gained acceptance, hospitals adopted the model for ED coverage. That, led to the development of ER patient treatment into a specialty of medicine.

He was invited, and became a charter member of the newly established American College of Emergency Physicians.

In 1992, after 20 years in solo practice, and with the advent of the health maintenance organizations (HMO) and other forms of managed-care organizations (MCO), Dr. Paguyo joined HealthPartners, Inc. —a staff model HMO where he retired in 2005, after 13 years with the company.

During his medical career, he took several courses in medical hypnosis, acupuncture, stress and chronic pain management, and nutrition, which he readily incorporated in his practice. He was a member of the medical staff at various Minneapolis/St. Paul area hospitals and chaired different hospital committees. At one time or another, he was a member of the AMA, Minnesota Medical Association, St. Paul and Hennepin County Medical Societies, and other subspecialty organizations in medicine. In addition, he was actively involved in the community; became a member of the Jaycees and Kiwanis, and Filipino organizations in the Twin Cities, especially the Fil-Minnesotan Association; twice elected president and member of the board for over 20 years.

He published in 2007 HEALTHCARE FOR ALL AMERICANS: Healthcare Crisis USA—A Comprehensive Solution. BETTER THAN OBAMACARE is the re-titled 2nd edition. He also wrote an unpublished but copyrighted treatise on universal identification (ID). This ID system was effectively demonstrated in the section on national healthcare ID card.

Dr. Nelson A. Paguyo is a dedicated and committed advocate for healthcare reform.

ABOUT THE BOOK

BETTER THAN OBAMACARE is the re-titled 2nd edition of HEALTHCARE FOR ALL AMERICANS: Healthcare Crisis USA— A Comprehensive Solution published in 2007.

It proposes ideal healthcare that is universal; comprehensive, portable, user-friendly, worry-free, reliable, uninterrupted in any situation an American may find himself/herself; simple to administer, affordable and sustainable; based on free market principles— applicable and adaptable for other countries.

It is a historical review of the U.S. healthcare system from post-World War II to the present.

It is a comparative study of the seven best known national healthcare schemes— Canada, England, France, Germany, Japan, Sweden, Switzerland and ObamaCare.

It is an examination and evaluation of the thirteen dominant U.S. healthcare problems.

It is an analysis how the U.S. healthcare impacts the American people, U.S. economy, and America's global competitiveness.

BOOKREVIEW

EUGENIO A. PULMANO, MD is a retired physician who practiced internal medicine for 35 years in the State of New Jersey.

A lot of work, research thought, talent and discipline went into the writing of this book.

While I was leafing through the table of contents, the introduction and then delving deeply into the "meat" of the book, now and then these thoughts popped to my mind: he is "crazy", and what makes him tick as to undertake such a formidable task? I could only surmise that the author believed he was equal to the task and more than proved it.

Dr. Nelson A. Paguyo has correctly made the set of differential diagnoses (the fault lines) in what ails the American healthcare systems and made some prescriptions that hopefully will bring health to the health system, but more importantly, bring better health to lives of real people.

The explosive growth of medical knowledge and technology is common knowledge. Its impact on medical care and healthcare delivery make a complicated system even more so, extending way beyond its proper domain and into the societal, political and economic realms.

The United States has the best in medicine with the latest scientific knowledge and cutting-edge technology. Translating that into a positive difference on the lives of millions of people is far less than impressive or even acceptable. If you factor in cost the results are even more appalling, especially when you make a comparative study with other developed nations which spend one-half or two-thirds of that spent in the US and get far better results.

One can thus characterize the US healthcare system as inefficient, ineffective and costly. It is dysfunctional.

When there are forty six to fifty million people who have no health insurance coverage, and thus virtually no access to health care except through the Emergency Department, which is the most expensive way to get medical care; not to mention is ineffective, inefficient, and virtually divorce from preventive care; when about an equal number are under insured and, therefore, out of pocket expenses from the patients are tremendous strain on their finances, it is no exaggeration to say the healthcare system is in crisis.

Out of genuine concern for millions of sick people and to bring some rationality to the system, Dr. Nelson Paguyo, retired pulmonologist felt impelled to undertake a study of the US health system with the idea of offering meaningful solutions.

His book, "Healthcare for All Americans: Healthcare Crisis USA--A Comprehensive Solution," in its second edition, entailed prodigious amount of work—library research, attendance at conferences, analysis, writing, and hours spent.

As healthcare is an exceedingly complex issue—just look back at the debates about it and ObamaCare year and tens of hundred or thousands of articles about it in old and new media to appreciate its humongous magnitude—I just like to focus on the core ideas of the book: solutions offered by Dr. Paguyo after his making comparative studies of healthcare all over the world, study of the various facets of the healthcare crisis in the US and assessment of ObamaCare, its plus and minuses.

He calls his proposed health system USA Universal Healthcare Plan (U.S.UHP). At the heart of it are two programs/agencies:

1.   Health Security Fund

2.   Health Security Agency

The Security Fund is the agency tasked to collect ALL health fund from all sources: Government and all its various agencies and programs—monies from Medicare, Medicaid, armed forces and veterans medical benefit programs, healthcare benefit program for federal employees, the members of the legislative, judicial and executive branches of he federal government, other federal entitlement programs, from the State for its medical assistance program; from the Private Sector including virtually all companies, corporations, small businesses, employees, self-employed individuals, retirees (married or not) except for those who are considered poor or unemployed. This Health Security Fund (HSF) is done at the national or federal level but replicated, coordinated and integrated in the state level.

The Health Security Agency (HSA) manages, administers, and disburses the Healthcare Security Fund's fund to all the states of the United States of America. Its secondary functions are to write (healthcare guidelines stated as general principles) for United States of America Universal Health Plan (with the help and collaborations of the Department of Health and Human Services, practicing medical experts, consumer advocates and other healthcare policy makers). It has other peripheral roles.

Now although, these agencies are national or federal in nature and are in some fashion replicated in the state level, many of the implementations of the policies and programs are done in collaboration with the private sector, especially with the health insurance companies. Indeed, the various activities are structured that way so as to put the private sector in better control of the health system, and not the public sector. The idea is to let the market has the upper hand, cognizant of the inefficiencies and deficiencies of many government programs.

The changes proposed are nothing short of radical: It will dismantle much of the healthcare system as we know it, and even as the current system is already incrementally transformed by ObamaCare. I suppose it is not bad to dismantle the system if what replaces has a good chance to work...for the better for the issues of access, cost and quality.

If the "cure" offered by Dr. Nelson Paguyo proves even two-thirds of what ails the US healthcare system is achievable, he would have done a Herculean job to alleviate the health needs of the people of the USA. All the labor and time he spent on it is all worth it.

Eugenio A. Pulmano, MD
Retired Internist, New Jersey

MR. JAMES K. MALONE is a reader of the book HEALTHCARE FOR ALL AMERICANS: Healthcare Crisis U.S.A. – A Comprehensive Solution.

Healthcare for all Americans by Dr. Nelson A. Paguyo is an extremely well written treatise on the healthcare crisis in this United States. He covers all aspects of the current health care system and the importance of their collective effects in contributing to a bright future of the health care for all Americans.

James K. Malone
Twin Cities, Minnesota